DVC

NELSON
CENGAGE Learning®

DVC: A journey from brief to design resolution
1st Edition
Paul Bourdot

Cover design: Cheryl Smith, McCarn Design
Text designer: Cheryl Smith, McCarn Design
Typset: Cheryl Smith, McCarn Design
Production controller: Siew Han Ong

Any URLs contained in this publication were checked for currency during the production process. Note, however, that the publisher cannot vouch for the ongoing currency of URLs.

© Cengage Learning Australia Pty Limited

For product information and technology assistance,
in Australia call **1300 790 853**;
in New Zealand call **0800 449 725**

For permission to use material from this text or product, please email
aust.permissions@cengage.com

National Library of New Zealand Cataloguing-in-Publication Data
A catalogue record for this book is available from the National Library of New Zealand.

ISBN 978 017 035557 5

Cengage Learning Australia
Level 7, 80 Dorcas Street
South Melbourne, Victoria, Australia 3205

Cengage Learning New Zealand
Unit 4B Rosedale Office Park
331 Rosedale Road, Albany, North Shore 0632, NZ

For learning solutions, visit **cengage.com.au**

Printed in China by China Translation & Printing Services.
1 2 3 4 5 6 7 18 17 16 15 14

DVC

A visual
journey from
brief to design resolution

Paul Bourdot

NELSON
CENGAGE Learning

Australia • Brazil • Japan • Korea • Mexico • Singapore • Spain • United Kingdom • United States

CONTENTS

CURRICULUM LEVEL 5

CURRICULUM LEVEL 6

PREFACE

This book has been prepared as a visual document to showcase student work and to provide an insight into the background skills needed to resolve each stage of a design solution, on the journey towards a completed portfolio.

The book deconstructs a design brief, by providing explanations and exercises that lead to the completion of each page of a portfolio. At the same time, the work provides examples of evidence that can be used for assessment purposes. It is intended to be used by both teacher and student, providing a resource of actual student work as well as the skill-building exercises, and their answers, that precede each stage.

Examples of student work at a low, middle and top level of achievement in design sketching, rendering and instrumental drawing are also included.

A unique feature is comment from the students whose work is profiled. Offering valuable insight into how they overcame the challenges posed by the brief, and the dedication required to produce the work, their words should be of value to all Design and Visual Communication students.

The book is aimed at a middle stage of a DVC course, curriculum levels 5 and 6, and is based on the Learning Objectives of the New Zealand curriculum:

- **Knowledge of design** — *students must demonstrate understanding of basic concepts in design.*
- **Human factors in design** — *students must demonstrate understanding of basic concepts and techniques related to humans factors in design.*
- **Visual communication** — *students must demonstrate skills in understanding of fundamental drawing techniques to present visual information.*
- **Graphics practice** — *students must demonstrate ability to apply design knowledge and drawing techniques to communicate design ideas.*

Students may need to refer back to my first book (*Year 9 Graphics*) for details on 'how to' for some of the foundation skills.

I wish to thank Lesley Pearce for her advice and encouragement, and the following students who have allowed me to use their work in this book. Without their passion, commitment and talent for design and visual communication, this resource would not have been possible.

Year 10
Shane du Plessis, James Morrison, Caroline Webb and Taylor Plank.

Year 11
Belinda Van Eeden, Nicole Dealey, Nathan Yee, Anton Weatherly, Mackenzie Farrell and Stephan Gailer.

Paul Bourdot
Paul is head of the Technology Faculty at Long Bay College in Auckland, specialising in Design and Visual Communication.

His students consistently gain top awards in competitions and the highest grades for assessments.

ISBN: 9780170355575

The teacher

For the successful delivery of a DVC programme, the classroom needs to be a well-considered space that works for both teacher and student.

It must be exciting and inviting, and allow for the best possible communication of visual information from the teacher, to ensure the best possible learning outcomes for the student. By taking advantage of readily available technology, the DVC classroom can become just that.

The advent of digital technology makes this easy. A visualiser is an essential aid in the DVC classroom. It consists of a flat working surface that can be lit from underneath (making use of 'old school' transparencies) or from above. A camera with zoom, tilt, freeze and a range of other options delivers a clear image from the work surface to a projector and onto a screen. Images can also be saved to a computer.

For reference while working, a monitor can be attached.

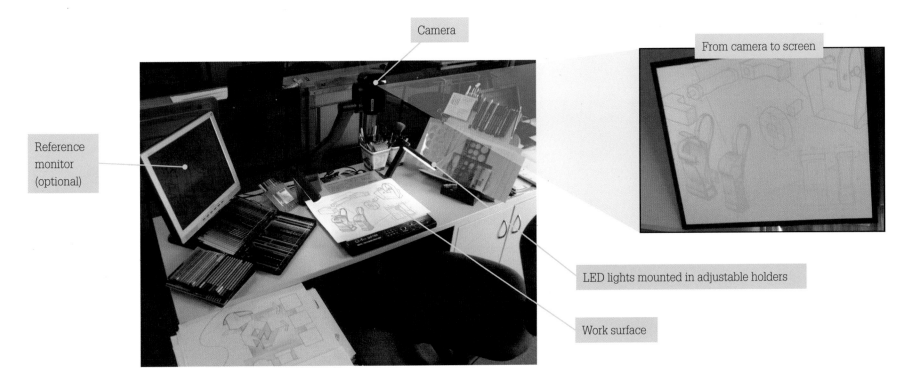

Camera

From camera to screen

Reference monitor (optional)

LED lights mounted in adjustable holders

Work surface

The student

Students should be made to feel special. Their workspaces should consist of the following.

- A comfortable chair with a backrest.
- An A3 drawing board, stored in a cradle at the front of the work space when not in use (A). This allows the flat part of the desk to be used for sketching, rendering, and model making, etc., preserving the drawing board for instrumental work only. A groove along the bottom of the board is provided for students to place their pencils.
- A T-square. When not in use it is stored on a hook at the end of the desk.
- A 'pull out' pencil-sharpening drawer (B). Consists of a replaceable strip of fine sandpaper, a strip of plain paper, and a pad of carpet. It is used to keep the point of the pencil sharp while working, to ensure line consistency.
- A stand for the placement of worksheets to keep the work surface tidy and to allow room for equipment to be laid out (C).
- A quick reference chart, attached to the desktop, that shows line types and printing standards (D).
- The drawing board cradle doubles as a place to store an equipment kit when the drawing board is being used (A).

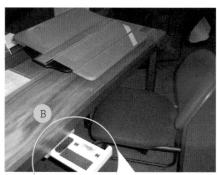

Access to computers and drawing tablets, and industry-standard software, is essential for developing digital literacy and future pathways.

The importance of student motivation, and the lifting of presentation skills using this technology, must not be overlooked.

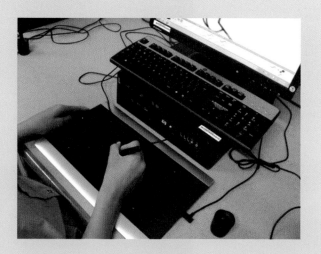

ISBN: 9780170355575

PROGRAMME MANAGEMENT

The indepth delivery of DVC makes demands on a teacher. The teaching of it cannot be 'bluffed'. Teachers must be skilled practitioners and be prepared to upskill at every opportunity. They must stay abreast of new techniques, new types of media and advances in technology.

To facilitate the delivery of a comprehensive programme of work, and to ensure the best possible outcomes for the student, careful management of time is essential.

There are two areas of study: product design and spatial design. Within each of these contexts, course content can be roughly divided into the following areas:

- Design and design sketching (ideation)
- Designers and design eras
- Presentation drawings

- Instrumental drawings (2D and 3D — orthographic projection, paraline drawing, perspective drawing).

Although teachers can 'pick and choose' from these areas to provide a design brief from which students will work, design sketching and presentation drawings need to be a compulsory element. Before work on the brief begins, skill-building exercises, directed by the teacher, must precede each key stage of the brief.

After each stage of the brief is completed, formative assessment needs to take place. Summative assessment can occur at a later date for the completed portfolio, after any amendments have been made by the student. The format below is suggested. Because all providers vary in their time allocation, no exact duration for tasks is given.

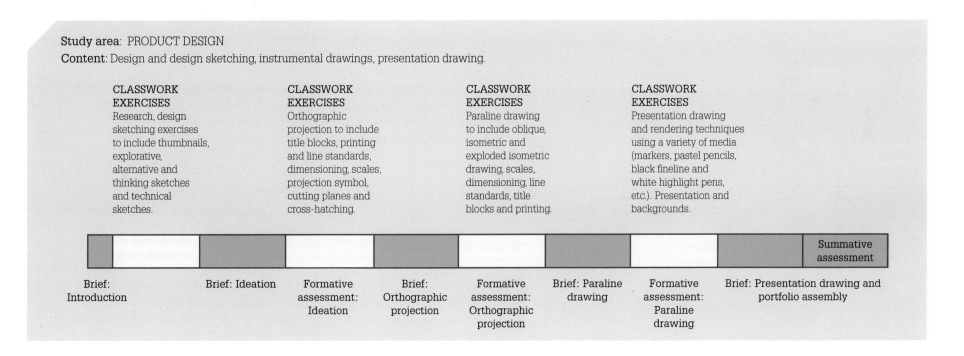

Study area: PRODUCT DESIGN

Content: Design and design sketching, instrumental drawings, presentation drawing.

CLASSWORK EXERCISES	CLASSWORK EXERCISES	CLASSWORK EXERCISES	CLASSWORK EXERCISES
Research, design sketching exercises to include thumbnails, explorative, alternative and thinking sketches and technical sketches.	Orthographic projection to include title blocks, printing and line standards, dimensioning, scales, projection symbol, cutting planes and cross-hatching.	Paraline drawing to include oblique, isometric and exploded isometric drawing, scales, dimensioning, line standards, title blocks and printing.	Presentation drawing and rendering techniques using a variety of media (markers, pastel pencils, black fineline and white highlight pens, etc.). Presentation and backgrounds.

Brief: Introduction | Brief: Ideation | Formative assessment: Ideation | Brief: Orthographic projection | Formative assessment: Orthographic projection | Brief: Paraline drawing | Formative assessment: Paraline drawing | Brief: Presentation drawing and portfolio assembly | Summative assessment

SPATIAL DESIGN

SITUATION
Holiday homes have become an integral part of the New Zealand landscape, where many people either own their own or rent to enjoy their holiday breaks.

BRIEF: Design a holiday bach suitable for a family.
DURATION: Six weeks (class time)

REQUIREMENTS

1. Design research. *Cut and pasted among your sketches.*
2. A bubble chart to show planning of the interior.
3. A line diagram to show the footprint of the design.

4. Ideation of the exterior of the bach. *Three 3D sketches — indicate your chosen design.*
5. Place notes where necessary, about your research and sketches, to explain your thinking and details that are not clear visually. Discuss *aesthetics* and *function.*
6. Technical sketches. *Develop your chosen design to show some construction details.*
7. Discuss your choice of building materials and sustainability.
8. A site plan. *Drawn to scale using instruments, to include the floor plan and landscape features.*
9. Elevations. *A minimum of two views, with labels indicating their relation to north.*
10. A rendered presentation drawing.

DESIGN SPECIFICATIONS
The bach must:
- Be single level.
- Accommodate up to 10 people.
- Have open plan living areas and have bedrooms, bathroom, laundry, office, a pool and a recreation area.
- Have a deck that offers easy access and indoor-outdoor flow.
- Take advantage of any natural environmental features (aspect).
- Take advantage of natural light (windows and doors).
- Consider aesthetics and function.
  Function: durability, efficiency, fitness for purpose, construction, user friendliness, safety, strength, size, scale, materials, etc.
  Aesthetics: style, proportion, harmony, form, rhythm, contrast, balance, texture, etc.

Planning

BUBBLE DIAGRAM

Here is an example of a simple spatial design brief. Introduced in the first quarter in the year, it has also been written to give students ready entry into external competitions.

It is important to note that no part of Shane's work, or any student work shown in this book, is a 'coincidence'. Every part has been taught using preceding skill-building exercises. For this brief, a practice brief for a container home, directed by the teacher, provided students with the skills and techniques they needed to work on their individual design briefs later (see 'Links to' below).

To achieve a consistent look to each page, a planning page was used. On this, students can decide on layout and colour palette, and make notes about what needs to be placed on each page.

Instead of 'scratchy' sketched bubbles, the use of a circle template, and on Shane's drawing, an ellipse template, smartens up the drawing. On the mock brief, students need to be shown techniques for rendering the bubbles, how to use a variety of media, and how to draw and render simple landscape features in plan view. Printing styles, neatness and accuracy for writing notes on the drawing, and for the titles, also need to be practised.

A good idea, as seen in Shane's drawing, is to write down the specs so that none is overlooked in the planning stages. Note the use of craft paper to add visual interest. Research has been found on the internet, then printed directly onto the page being used.

Note also the subtle use of smudged pastel pencil to create a background.

 SKILLS NEEDED TO PRECEDE THIS PAGE:

Planning sheet.
Drawing and rendering techniques for bubbles, arrows, 2D landscape graphics.
Research and simple design notes that explain thinking not clear visually.
Media technique: marker pens, black fineline pens, pastel pencils, colouring pencils.
Links to: pages 27–29, 34–39, 77

ISBN: 9780170355575   PHOTOCOPYING OF THIS PAGE IS RESTRICTED UNDER LAW.

Well balanced and has a unique harmony. Great indoor/outdoor flow extending your living area onto the deck.

Wooden cladding gives a nice texture to the outside and adds to the finish. Very efficiently spaced to give an open living plan.

Infinity pool adds to the rhythm of the layout. Extending your living area into the ocean.

It uses space effectively and adds to the finish of the whole design. A well constructed design.

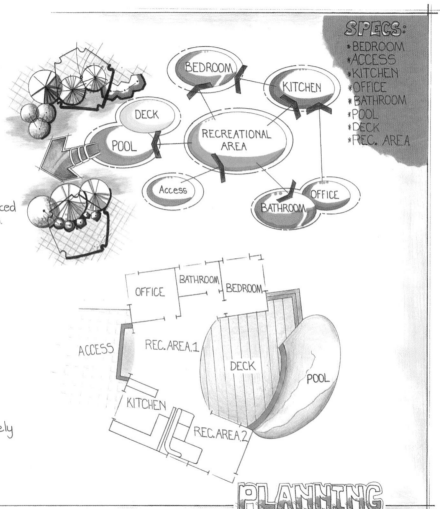

SPECS:
* BEDROOM
* ACCESS
* KITCHEN
* OFFICE
* BATHROOM
* POOL
* DECK
* REC. AREA

BEDROOM

KITCHEN

DECK

RECREATIONAL AREA

POOL

Access

OFFICE

BATHROOM

OFFICE | BATHROOM | BEDROOM

ACCESS

REC. AREA. 1

DECK

POOL

KITCHEN

REC. AREA. 2

PLANNING

This page took a lot of time. The challenging part was drawing all the bubbles and thinking about how the rooms would connect.

I did a practice page first to make sure I had a good page layout, then got some research from the internet. Because the DVC department at school had an A3 colour printer, I printed this onto the A3 page then added my drawings.

I drew the bubbles with an ellipse template then outlined them with a 0.3 and 0.1 black fineline pen. The rendering was done with markers. I practised printing first, to make it look neat in the bubbles, around my research and in the line diagram.

I researched how to draw the plants, which were not as hard as I thought they would be. I used a circle template and then just added interesting shapes and rendered them with marker pens. I like they way they turned out.

Overall I am happy with this page. If I had to do it again I would change the layout of the research.

SPATIAL DESIGN

SITUATION

Holiday homes have become an integral part of the New Zealand landscape, where many people either own their own or rent to enjoy their holiday breaks.

BRIEF: Design a holiday bach suitable for a family.
DURATION: Six weeks (class time)

REQUIREMENTS

1 Design research. *Cut and pasted among your sketches.*
2 A bubble chart to show planning of the interior.
3 A line diagram to show the footprint of the design.

4 Ideation of the exterior of the bach. *Three 3D sketches — indicate your chosen design.*
5 Place notes where necessary, about your research and sketches, to explain your thinking and details that are not clear visually. Discuss *aesthetics* and *function*.

6 Technical sketches. *Develop your chosen design to show some construction details.*
7 Discuss your choice of building materials and sustainability.
8 A site plan. *Drawn to scale using instruments, to include the floor plan and landscape features.*
9 Elevations. *A minimum of two views, with labels indicating their relation to north.*
10 A rendered presentation drawing.

DESIGN SPECIFICATIONS

The bach must:
- Be single level.
- Accommodate up to 10 people.
- Have open plan living areas and have bedrooms, bathroom, laundry, office, a pool and a recreation area.
- Have a deck that offers easy access and indoor-outdoor flow.
- Take advantage of any natural environmental features (aspect).
- Take advantage of natural light (windows and doors).
- Consider aesthetics and function.
  Function: durability, efficiency, fitness for purpose, construction, user friendliness, safety, strength, size, scale, materials, etc.
  Aesthetics: style, proportion, harmony, form, rhythm, contrast, balance, texture, etc.

Ideation

The second page begins the exploration of ideas for the shape of the dwelling. Again, informed by research, three freehand sketches, preferably in perspective, were asked to be shown.

The layout of each has been planned on the planning sheet to ensure the best viewpoints and correct proportion.

The development of a sophisticated sketching style needs to be encouraged, along with the drawing and rendering of 3D landscape features, materials and sky.

The placement of the horizon line is a critical element to the success of Shane's sketches, helping to also highlight the design features.

Indication of the chosen design should feature on this page, along with notes to explain thinking that is not clear visually.

Note the careful attention to finishing the page with a smudged pastel background, and a border that ties the sketches together. The border also aids in drawing the eye to the flow of ideas.

Confident use of rendering media enhances each sketch, providing practice in skills and techniques that can be used on the final presentation drawing.

The hand-drawn title completes the page.

 SKILLS NEEDED TO PRECEDE THIS PAGE:

Planning sheet, freehand design sketching: perspective drawing in one and two point.
Viewpoints, crating (proportions) and line hierarchy.
Rendering techniques — light source, tonal changes, form, materials, landscape features, etc.
Research and simple design notes that explain thinking not clear visually.
Media technique: marker pens, black fineline pens, pastel pencils.
Links to: pages 22–26, 34–39, 77

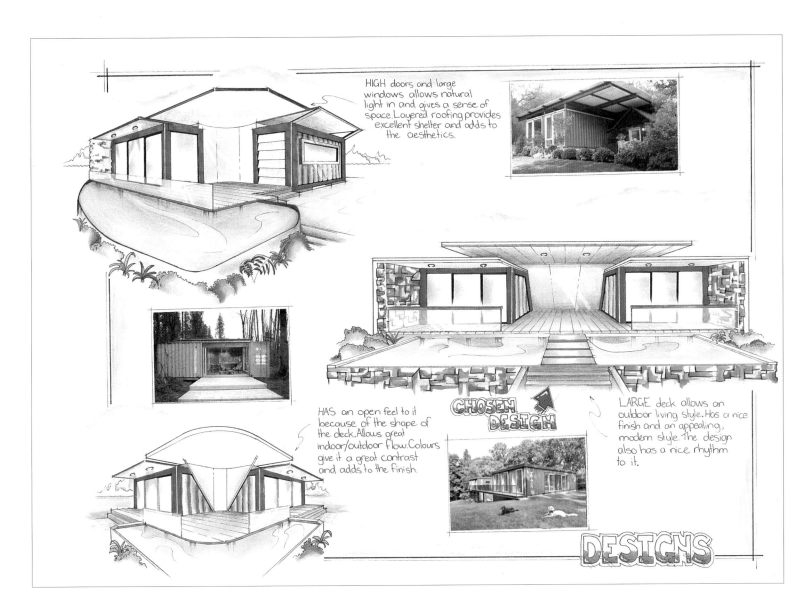

HIGH doors and large windows allows natural light in and gives a sense of space. Layered roofing provides excellent shelter and adds to the aesthetics.

HAS an open feel to it because of the shape of the deck. Allows great indoor/outdoor flow. Colours give it a great contrast and adds to the finish.

CHOSEN DESIGN

LARGE deck allows an outdoor living style. Has a nice finish and an appealing, modern style. The design also has a nice rhythm to it.

DESIGNS

This page took the longest time of all. But I am glad I spent the time because it looks good.

I did a practice page layout first to get the right viewpoints and make sure the views would fit.

I used perspective drawings and kept the horizon lines low to show off the walls more.

The part I found hard was rendering the different materials. I wanted to make sure that you can tell by looking at the drawings what they are made from. Showing the bricks on the middle drawing was hardest.

I used a black fineline pen to outline the drawings, then marker pens, colouring pencils and pastel pencils to render them. I liked the way the sky and the water turned out. I learned that by putting some vertical lines on water parts makes it look reflective. I tore a piece of paper and smudged blue pastel on the edge for the sky.

If I was to do the drawing again I would change the title block. I didn't leave enough room for it.

SPATIAL DESIGN

SITUATION

Holiday homes have become an integral part of the New Zealand landscape, where many people either own their own or rent to enjoy their holiday breaks.

BRIEF: Design a holiday bach suitable for a family.
DURATION: Six weeks (class time)

REQUIREMENTS

1 Design research. *Cut and pasted among your sketches.*
2 A bubble chart to show planning of the interior.
3 A line diagram to show the footprint of the design.
4 Ideation of the exterior of the bach. *Three 3D sketches — indicate your chosen design.*
5 Place notes where necessary, about your research and sketches, to explain your thinking and details that are not clear visually. Discuss *aesthetics* and *function.*

6 Technical sketches. *Develop your chosen design to show some construction details.*
7 Discuss your choice of building materials and sustainability.

8 A site plan. *Drawn to scale using instruments, to include the floor plan and landscape features.*
9 Elevations. *A minimum of two views, with labels indicating their relation to north.*
10 A rendered presentation drawing.

DESIGN SPECIFICATIONS

The bach must:
- Be single level.
- Accommodate up to 10 people.
- Have open plan living areas and have bedrooms, bathroom, laundry, office, a pool and a recreation area.
- Have a deck that offers easy access and indoor-outdoor flow.
- Take advantage of any natural environmental features (aspect).
- Take advantage of natural light (windows and doors).
- Consider aesthetics and function.
  Function: durability, efficiency, fitness for purpose, construction, user friendliness, safety, strength, size, scale, materials, etc.
  Aesthetics: style, proportion, harmony, form, rhythm, contrast, balance, texture, etc.

Ideation (cont.)

TECHNICAL SKETCHES

Once the final design has been chosen, it's time to consider how the design, or parts of the design, could be made.

Sometimes called *Design Development*, technical sketches should be a mix of 3D exploded isometric or oblique, and 2D sectioned views of parts.

Shane has shown this extremely well by keeping things simple and focusing on areas that match his understanding of construction. To enable this page to happen though, students need access to resource books and the internet.

Research needs to be ongoing. This page also demonstrates Shane's skill and knowledge of drawing systems, particularly 3D exploded and 2D sectioned views, also taught on the preceding 'mock' teacher-directed brief.

Again, the page layout has been planned to maintain the look and feel of the preceding pages.

Plenty of white space, a background and border, confident rendering and symbolism (arrows), title and neat printing of notes, lift the page to a higher level.

 SKILLS NEEDED TO PRECEDE THIS PAGE:

Planning sheet used. Colour palette — soft tones.
Freehand design sketching in 2D and 3D and line hierarchy.
Viewpoints are considered for visual interest and to better show the ideas.
Correct 'exploded' techniques. The parts are pulled apart in the direction in which they would be assembled.
Simple 2D cross-hatching technique.
Simple notes explain thinking not clear visually. Arrows indicate movement and/or function.
Rendering techniques — minimalist, indicating the light source, tonal changes, and form.
Links to: pages 34–39, 77 and resource books, internet

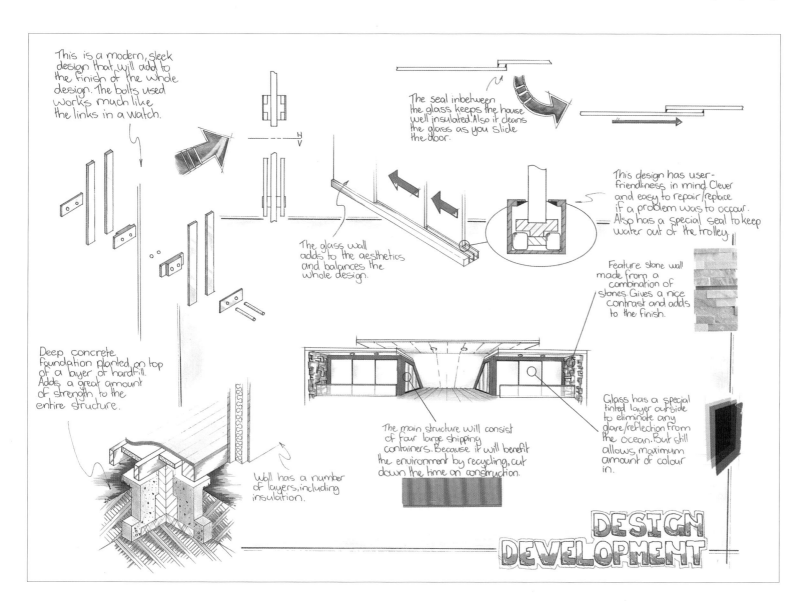

This is a modern, sleek design that will add to the finish of the whole design. The bolts used works much like the links in a watch.

The seal inbetween the glass keeps the house well insulated. Also it cleans the glares as you slide the door.

This design has user-friendliness in mind. Clever and easy to repair/replace if a problem was to occur. Also has a special seal to keep water out of the trolley.

The glass wall adds to the aesthetics and balances the whole design.

Feature stone wall made from a combination of stones. Gives a nice contrast and adds to the finish.

Deep concrete foundation planted on top of a layer of hardfill. Adds a great amount of strength to the entire structure.

Glass has a special tinted layer outside to eliminate any glare/reflection from the ocean. But still allows maximum amount of colour in.

The main structure will consist of four large shipping containers. Because it will benefit the environment by recycling, cut down the time on construction.

Wall has a number of layers, including insulation.

DESIGN DEVELOPMENT

I found this page a challenge because I needed to research how my design could be made.

I used the internet and books in my classroom to help, but then had to work out what kinds of drawings would be best to use.

The hardest one was the cutaway isometric view. I knew how to draw an exploded view because I had done these on my design assignment in Year 9.

I used marker pens to render, which was the same as my other page, so this was the easiest part.

When I finished, the page looked a bit empty so I added a background using smudged pastel to match the other pages.

Although I am quite happy with the way the drawings turned out, I should have made them bigger. This would make it easier to see some of the details.

SITUATION

Holiday homes have become an integral part of the New Zealand landscape, where many people either own their own or rent to enjoy their holiday breaks.

BRIEF: Design a holiday bach suitable for a family.
DURATION: Six weeks (class time)

REQUIREMENTS

1 Design research. *Cut and pasted among your sketches.*
2 A bubble chart to show planning of the interior.
3 A line diagram to show the footprint of the design.
4 Ideation of the exterior of the bach. *Three 3D sketches — indicate your chosen design.*
5 Place notes where necessary, about your research and sketches, to explain your thinking and details that are not clear visually. Discuss *aesthetics* and *function.*
6 Technical sketches. *Develop your chosen design to show some construction details.*
7 Discuss your choice of building materials and sustainability.

8 A site plan. *Drawn to scale using instruments, to include the floor plan and landscape features.*

9 Elevations. *A minimum of two views, with labels indicating their relation to north.*
10 A rendered presentation drawing.

DESIGN SPECIFICATIONS

The bach must:
- Be single level.
- Accommodate up to 10 people.
- Have open plan living areas and have bedrooms, bathroom, laundry, office, a pool and a recreation area.
- Have a deck that offers easy access and indoor-outdoor flow.
- Take advantage of any natural environmental features (aspect).
- Take advantage of natural light (windows and doors).
- Consider aesthetics and function.
  Function: durability, efficiency, fitness for purpose, construction, user friendliness, safety, strength, size, scale, materials, etc.
  Aesthetics: style, proportion, harmony, form, rhythm, contrast, balance, texture, etc.

Instrumental drawing

SITE AND FLOOR PLAN

Now is the opportunity to use a drawing board and instruments to accurately produce, to scale, the footprint floor plan of the dwelling.

Precision of line and skill when using instruments is clearly seen in Shane's drawing.

But again, students need to do a practice drawing beforehand, with emphasis on line quality.

Students need to be taught how to dimension architectural drawings, and have access to resources that show how to draw architectural symbols for the interior and how to draw a north symbol.

For this drawing, they should be encouraged to outline the work in ink, using 0.1 and 0.3 black fineline pens to draw architectural symbols.

Once outlined, the drawing could be cut and pasted onto another sheet.

Drawing selected landscape features and entourage detail such as the car, and rendering them with care, provides the page with the 'wow' factor.

 SKILLS NEEDED TO PRECEDE THIS PAGE:

Line weights and line standards (pencil to begin with, then inked in — 0.1 and 0.3 black fineline pens).
Title and printing standards.
Scales and the north symbol.
Correct dimensioning techniques for architectural drawing.
Drawing and rendering of landscape features and entourage detail.
Colour palette — soft tones.
Links to: pages 27–29, 77, 183

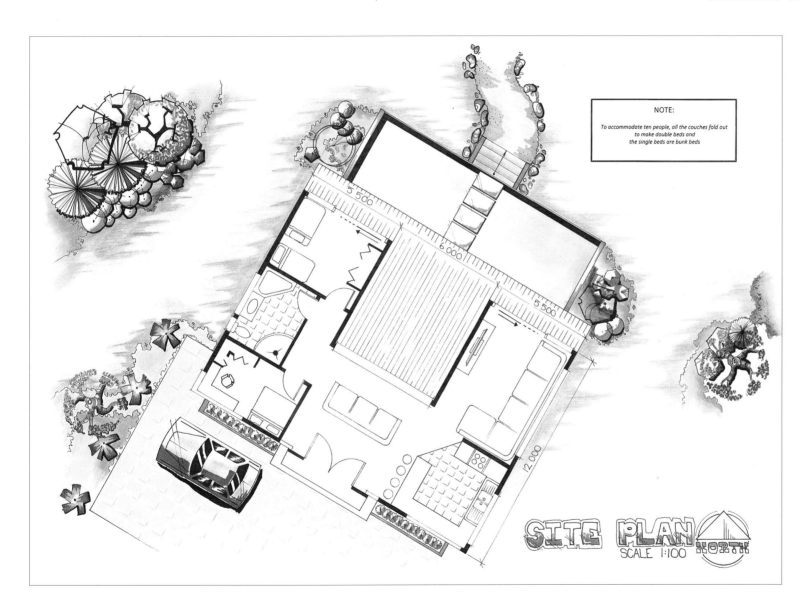

NOTE:

To accommodate ten people, all the couches fold out
to make double beds and
the single beds are bunk beds

5 500

6 000

5 500

12 000

SITE PLAN
SCALE 1:100   NORTH

The hardest part about this page was working out the scale so it would fit.

The best part was getting the look of the foliage right. I used books in my classroom that had some good examples. But I had to practise them first. I used a circle template to get the proportions right, then added the shapes.

I found that by not worrying about making the plants look real, by using interesting shapes and different colours, I could get a more stylish look.

I used marker pens, colouring pencils and pastels to render the page, and outlined all the parts with a black pen … I am really happy with the way this page looks, even the blue colouring pencil I used for the ground around the house.

I think the car looks a bit too big though.

SPATIAL DESIGN

SITUATION
Holiday homes have become an integral part of the New Zealand landscape, where many people either own their own or rent to enjoy their holiday breaks.

BRIEF: Design a holiday bach suitable for a family.
DURATION: Six weeks (class time)

REQUIREMENTS
1 Design research. *Cut and pasted among your sketches.*
2 A bubble chart to show planning of the interior.
3 A line diagram to show the footprint of the design.
4 Ideation of the exterior of the bach. *Three 3D sketches — indicate your chosen design.*
5 Place notes where necessary, about your research and sketches, to explain your thinking and details that are not clear visually. Discuss *aesthetics* and *function*.
6 Technical sketches. *Develop your chosen design to show some construction details.*
7 Discuss your choice of building materials and sustainability.
8 A site plan. *Drawn to scale using instruments, to include the floor plan and landscape features.*

9 Elevations. *A minimum of two views, with labels indicating their relation to north.*

10 A rendered presentation drawing.

DESIGN SPECIFICATIONS
The bach must:
- Be single level.
- Accommodate up to 10 people.
- Have open plan living areas and have bedrooms, bathroom, laundry, office, a pool and a recreation area.
- Have a deck that offers easy access and indoor-outdoor flow.
- Take advantage of any natural environmental features (aspect).
- Take advantage of natural light (windows and doors).
- Consider aesthetics and function.
  Function: durability, efficiency, fitness for purpose, construction, user friendliness, safety, strength, size, scale, materials, etc.
  Aesthetics: style, proportion, harmony, form, rhythm, contrast, balance, texture, etc.

Instrumental drawing

ELEVATIONS
Again, using a drawing board and instruments, and drawing to scale, selected elevations of the dwelling can be shown.

Precision of line and skill when using instruments is again a priority for this drawing. And again, students need to do a practice drawing beforehand.

They need access to resources that show good examples of how elevations can be enhanced by working in ink (pencil first though) and adding some 2D landscape features.

For this drawing, Shane has opted for a clean black and white look, enhanced by different pen thicknesses. The addition of a border that ties the work together, and the title with scale, completes the page well.

Note the neat printing (between guide lines) of labels of the views as they appear in relation to north.

 SKILLS NEEDED TO PRECEDE THIS PAGE:

Line weights and line standards (pencil to begin with, then inked in — 0.1, 0.3 and 0.5 black fineline pens).
Title and printing standards.
Scales.
Correct dimensioning techniques for architectural drawing.
Drawing of rendering of landscape features.
Colour palette — black and white.
Links to: pages 31, 34–39, 77

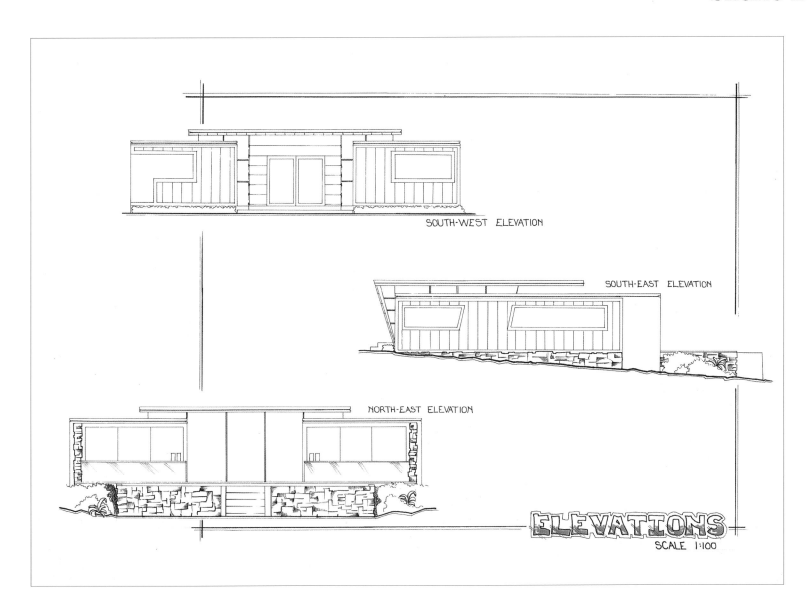

SOUTH-WEST ELEVATION

SOUTH-EAST ELEVATION

NORTH-EAST ELEVATION

ELEVATIONS

SCALE 1:100

The most difficult part of this page was working out the scale.

I wanted to show all the elevations on one page, so it took time to get the sizes right and not make the drawings too small.

Drawing the elevations was quite easy once I started. It took a lot of time to outline them in ink pen though. I had to make sure I didn't smudge the lines. I used a 0.1 and 0.3 black fineline pen but did them in pencil first.

I looked in some architecture books to get a professional look. I really like the double thick and thin lines under each view for the ground lines.

I was going to print a title on the computer and paste it on the drawing but decided to do it by hand to match my other pages. I printed the names of the views in ink. This took a while because I wanted to make it as neat as I could.

If I had more time I would have included one or two more landscape elements.

SITUATION

Holiday homes have become an integral part of the New Zealand landscape, where many people either own their own or rent to enjoy their holiday breaks.

BRIEF: Design a holiday bach suitable for a family.
DURATION: Six weeks (class time)

REQUIREMENTS

1 Design research. *Cut and pasted among your sketches.*
2 A bubble chart to show planning of the interior.
3 A line diagram to show the footprint of the design.
4 Ideation of the exterior of the bach. *Three 3D sketches — indicate your chosen design.*
5 Place notes where necessary, about your research and sketches, to explain your thinking and details that are not clear visually. Discuss *aesthetics* and *function*.
6 Technical sketches. *Develop your chosen design to show some construction details.*
7 Discuss your choice of building materials and sustainability.
8 A site plan. *Drawn to scale using instruments, to include the floor plan and landscape features.*
9 Elevations. *A minimum of two views, with labels indicating their relation to north.*

10 A rendered presentation drawing.

DESIGN SPECIFICATIONS

The bach must:
- Be single level.
- Accommodate up to 10 people.
- Have open plan living areas and have bedrooms, bathroom, laundry, office, a pool and a recreation area.
- Have a deck that offers easy access and indoor-outdoor flow.
- Take advantage of any natural environmental features (aspect).
- Take advantage of natural light (windows and doors).
- Consider aesthetics and function.
  Function: durability, efficiency, fitness for purpose, construction, user friendliness, safety, strength, size, scale, materials, etc.
  Aesthetics: style, proportion, harmony, form, rhythm, contrast, balance, texture, etc.

Presentation drawing

It's time now to showcase student talent by producing a realistic rendered drawing of the design. The best type of drawing system for spatial design is perspective, with the most important consideration being viewpoint, and the placement of the horizon line.

Shane's choice of one-point perspective offers a dramatic viewpoint that draws the eye into the design features. The drawing is large enough to facilitate rendering of all surfaces and features.

Once a satisfactory pencil drawing has been done, a good idea is to make several photocopies on cartridge paper (not the thin photocopy paper) to allow for the use of marker pens. Photocopying will also enable the drawing to be enlarged or reduced to suit. Preserve one copy as the master, on which the final rendering can be done, and use the others for practice. Once the master is fully rendered and outlined, landscape features can be drawn.

Shane has used marker pens, pastel pencil for the sky, windows and water, and colouring pencils for the grass and distant landscape features. The drawing has been outlined in black fineline ink pen. Note the effective use of shadows — the roof overhang onto the walls and windows, on the steps leading up to the pool area, and reflection in the pool. The addition of these features play an important role in adding to the three-dimensionality of the drawing.

Remember, though, that these techniques must be taught — they are not a given.

Precede this drawing with exercises to give practice in applying a range of media to different surface textures and landscape features, which will ultimately inform student decisions for their own design.

SKILLS NEEDED TO PRECEDE THIS PAGE:

Perspective drawing techniques, viewpoint and backgrounds.
Crating (proportions) and line hierarchy.
Rendering techniques — light source, tonal changes, shape and surface texture.
Colour palette considered.
Media technique: marker pens, black fineline pens, pastel pencils.
Links to: pages 22–26, 30–33, 34–39, 77

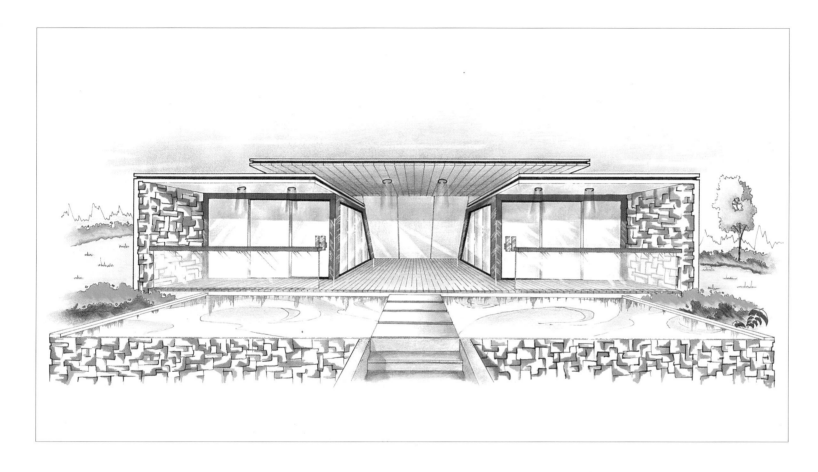

This drawing is basically a larger version of the one on my ideation page, so I didn't find it too hard to do.

To get a good viewpoint, I used one-point perspective and like how it gives a view of walking up to the building.

What was hard was picking a suitable background. I kept it simple and even though I rushed the tree a bit, I think my landscape works well.

Because I had already done a lot of similar rendering on other drawings in class, I used the same techniques on this drawing, using mainly marker pens. I used colouring pencils for some of the background landscape and outlined the whole drawing with a thin black pen.

I especially like the effects I got of the shadow of the roof on the walls by smudging 4B pencil, and the reflection in the pool.

I should have spent more time on the foliage around the pool to make it look better.

PERSPECTIVE DRAWING

Introduction

One of the main components of a spatial drawing is a perspective drawing.

When you take a photograph of an object, the resulting image will be in perspective. As in a photograph, perspective drawings look realistic because objects appear to get smaller the further away they are. The lines of the object also appear to narrow to a point in the distance called the **vanishing point** (VP).

The vanishing point is on the **horizon**, also called the **eye level** line. Lines that travel to the vanishing point are called **visual rays**.

There are two types of perspective drawing:

One-point perspective — one vanishing point (VP), the object being viewed looking at the front or side.

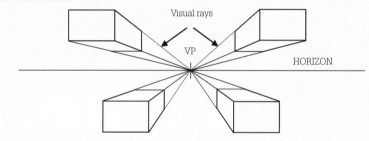

Two-point perspective — two vanishing points (VP1, VP2), the object being viewed looking at a corner or edge.

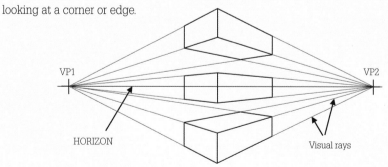

A main skill of perspective drawing is to divide a distance into equal spaces, to compensate for the foreshortening that occurs as the spaces go further away from you.

1 Finding equal spaces when given the distance between each space.

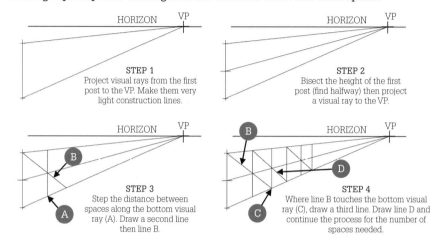

STEP 1
Project visual rays from the first post to the VP. Make them very light construction lines.

STEP 2
Bisect the height of the first post (find halfway) then project a visual ray to the VP.

STEP 3
Step the distance between spaces along the bottom visual ray (A). Draw a second line then line B.

STEP 4
Where line B touches the bottom visual ray (C), draw a third line. Draw line D and continue the process for the number of spaces needed.

2 Finding equal spaces when you know the length of the space.

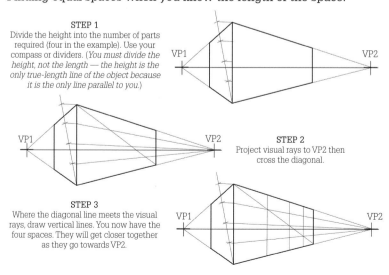

STEP 1
Divide the height into the number of parts required (four in the example). Use your compass or dividers. (*You must divide the height, not the length — the height is the only true-length line of the object because it is the only line parallel to you.*)

STEP 2
Project visual rays to VP2 then cross the diagonal.

STEP 3
Where the diagonal line meets the visual rays, draw vertical lines. You now have the four spaces. They will get closer together as they go towards VP2.

ISBN: 9780170355575

Circles

Circles and cylinders in perspective can be sketched freehand by placing them inside a square crate. The same principle as used for a sketched 2D circle is used:

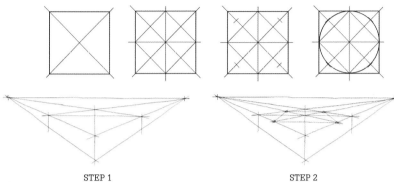

STEP 1
Sketch or use instruments to draw the perspective crate. Cross the diagonals to find the centre.

STEP 2
Where the diagonals intersect, draw the centre lines to the vanishing points.

STEP 3
Find the halfway point between the diagonal line and the corner of the crate.

STEP 4
Carefully sketch the curve to pass through the points on the crate where the centre lines touch. Add depth if it is a cylinder.

Shadows

You can add a shadow to a curved object.

1 Establish plotting points on the edge of the top curve, then project them vertically to the bottom curve.

2 Establish the light source (LS) and shadow vanishing point (SVP) directly below the light source.

3 Project visual rays from the light source and SVP to form the shadow.

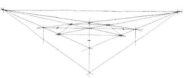

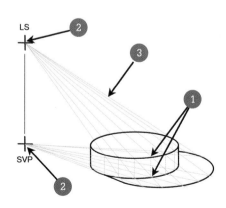

Shadows (cont.)

Placing shadows on your perspective drawings will make them more realistic. Shadows are especially effective on architectural drawings and on objects that reflect onto shiny surfaces. Shadows can be constructed either with instruments or freehand.

Steps for drawing a shadow of a cube, with instruments, in two-point perspective:

1 Draw the horizon line and vanishing points. Bisect the horizon and open your compass from this to VP1. Draw a semi-circle between VP1 and VP2. This arc represents the path of the sun from one point on the horizon to the next. *The vanishing point for the shadow will be on the horizon and the light source will be on the arc.*

2 Place the light source (LS) anywhere on the curve and extend it vertically to the horizon line. This point becomes the shadow vanishing point (SVP).

3 Draw lines (visual rays) from the light source through the top corners of the object. These represent light rays.

4 Draw lines from the SVP through the bottom corners of the object to intersect with those from the light source. These lines are the direction of light.

5 Join the outlines to form the shadow. Render the shadow the same tone as the darkest surface of the object.

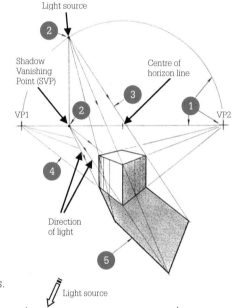

You can also add shadows to freehand sketches as shown in these simple steps.

STEP 1
Sketch the object including hidden edges.

STEP 2
Draw a triangle formed by the angle from the light source and the direction of light.

STEP 3
Duplicate the shadow triangle from all sides with parallel lines.

STEP 4
Connect the bases of the triangles to form the shadow.

STEP 5
Render the shadow a darker tone of the surface on which it falls.

Introduction exercises

Draw this on an A3 page, then do it as an introduction to develop the following perspective drawing skills.

- Line types — construction (visual rays), outlines.
- How to draw a perspective circle, how to divide spaces equally, how to draw a shadow.
- How to draw in one-point (parallel) perspective and two-point (angular) perspective.
- How to draw the horizon line and vanishing points.

EXERCISE 1a
In one-point perspective, construct five equally spaced lampposts when the distance between each is given.
1   Draw the horizon line and vanishing point.
2   Locate and draw the first lamppost shown. (Judge the height and size of the crossbar.)
3   Draw visual rays to the vanishing points.
4   Draw the remaining four posts using the construction method for equal spaces.

EXERCISE 1b
In two-point perspective, construct the cube and project its shadow from the given light source.
1   Draw the horizon line and vanishing points.
2   Locate and draw the front corner of the cube.
3   Draw visual rays to the vanishing points, then draw the back corners and then the cube.

4   Locate the light source and draw the shadow.

EXERCISE 1c
In two-point perspective, redraw the cabinet shown with three equally spaced drawers and two equally spaced cupboards. Show the two walls behind the cabinet.
1   Draw the horizon line and vanishing points.
2   Draw the front corner of the cabinet.
3   Draw visual rays to the vanishing points.
4   Measure along the top visual rays the size of the cabinet from the measurements in the plan.
5   Draw the perspective box to contain the cabinet.
6   Draw the base of the cabinet. (Judge the sizes.)
7   Construct the equally spaced doors and drawers, add the walls and any other detail such as a floor-standing lamp.

EXERCISE 1a

In one-point perspective, construct five equally spaced lamp-posts when the distance between each is given.

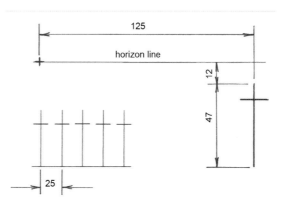

EXERCISE 1c

In two-point perspective, redraw the cabinet shown with three equally spaced drawers and two equally spaced cupboards.

Show the two walls that the cabinet is placed against.

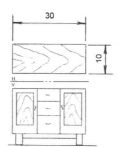

EXERCISE 1b

In two-point perspective, construct the cube and project its shadow from the given light source.

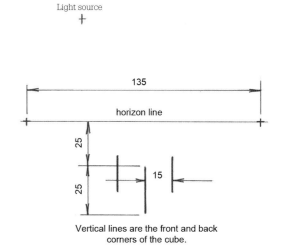

Vertical lines are the front and back corners of the cube.

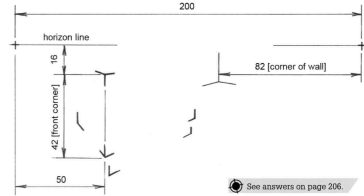

See answers on page 206.

ISBN: 9780170355575

Do these exercises to develop the following perspective drawing skills:

- Line types — construction (visual rays), outlines.
- How to draw circles and curves in perspective.
- How to draw in one-point (parallel) perspective and two-point (angular) perspective.
- How to draw the horizon line and vanishing points.
- How to divide a length into equal spaces.
- How to divide a given distance into equal spaces.

EXERCISE 2: BEDROOM

1   Follow the numbered steps to carefully redraw the layout shown below, onto an A3 sheet.
2   Work from the floor plan shown on the right to draw the one-point perspective view.
3   Show all constructions clearly and render the finished drawing.

Rendering media: *pastel pencil, colouring pencil, black fineline pen, marker pen.*

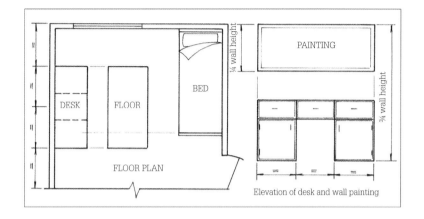

FLOOR PLAN

Elevation of desk and wall painting

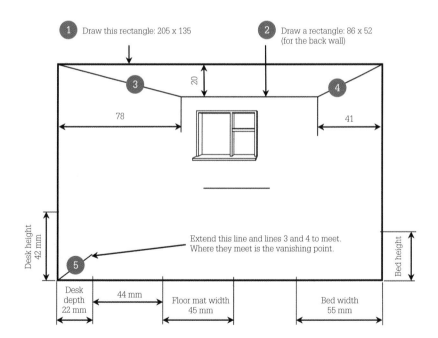

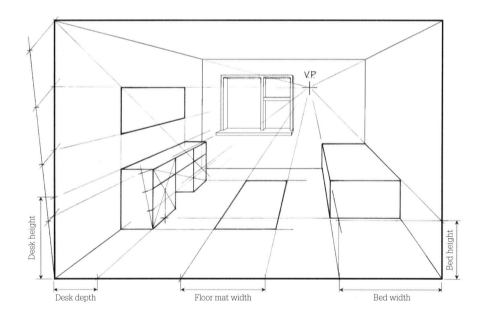

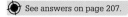
See answers on page 207.

EXERCISE 3: HEATER

1 Follow the numbered steps to carefully redraw the layout shown below, onto an A3 sheet. *Sizes are from the paper edge.*

2 Working from the given sizes, and looking closely at the orthographic projection shown, redraw the heater in two-point perspective. *Judge any sizes not given.*

3 Show all constructions clearly and render the finished drawing.

See answers on page 207.

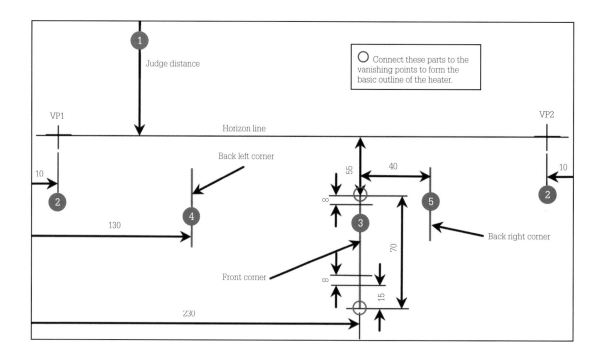

Judge distance

Connect these parts to the vanishing points to form the basic outline of the heater.

VP1   Horizon line   VP2

10   Back left corner   55   40   10

2   130   4   8   3   70   5   Back right corner

Front corner   8

230   15

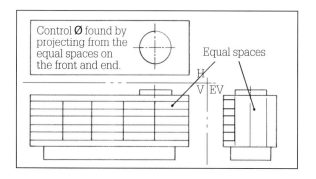

Control Ø found by projecting from the equal spaces on the front and end.

Equal spaces

H
V EV

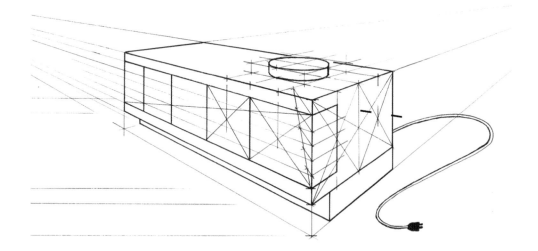

ISBN: 9780170355575

Bubble diagrams

A designer or architect needs to plan how all the rooms or spaces of a building will connect with each other, before drawings of the building can begin. This is done by using a 'bubble diagram'.

A bubble diagram is a simple drawing of roughly drawn bubbles (representing spaces) connected by arrows, broken lines or wavy lines, etc. to show how the spaces connect with each other. For example, a lounge area may need to connect to the dining room with a visual connection to outside views of the ocean and lead out to a deck.

A good place to begin is to make a list of all the rooms and spaces needed, then place them in bubble shapes. Link the 'bubbles' to each other with arrows.

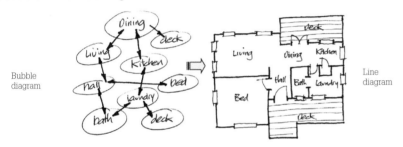

Bubble diagram

Line diagram

A bubble diagram for a simple one-bedroom house is shown above left. When the final bubble arrangement has been chosen, a line sketch (line diagram) can be drawn (above right), working towards the 'footprint shape' or floor plan.

You can make a feature of the bubbles by drawing an outside broken line with a black pen then rendering them with markers. Use a circle or ellipse template for an even neater look. 'Styley' printing and more interesting arrow shapes add visual appeal.

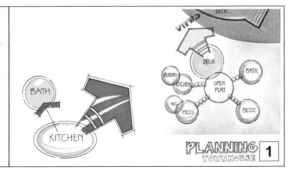

Floor plans

Floor plans are drawn to scale, using instruments or a computer.

Floor plans may show some or all of the following: *dimensions, electrical wiring symbols, walls, door and window openings, features in the rooms such as bath, toilet, bed, kitchen bench, rooms labelled and sometimes with their dimensions, decks, stairways, etc.*

The simple floor plan of a house shows conventions for drawing walls, doors, windows and dimensions. The dashed line shows the overhang of the roof. *For clarity, interior walls have not been shown on this drawing.*

Note that the dimension line may have an angled line instead of an arrowhead.

Wall thickness is usually 100 mm and may be shaded in black with an ink pen.

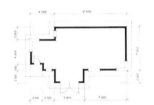

The floor plan of a single-car garage shows the following electrical symbols: *water tap, fluorescent lights, light switches, wall light, doorway, roller shutter door, and window.*

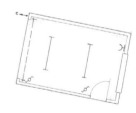

The drawing at left is typical of that drawn by an architect. Rooms have been named with some fittings shown. Note how lines have been allowed to overhang at the corners.

Site plans

Site plans show the detail of the land on which a building is placed. The floor plan of the building may also be shown. The site plan is drawn to scale with north indicated. Landscape features and rendering enhances the visual appearance of the drawing.

NORTH SYMBOLS
A north symbol is placed on the site plan to show the direction the site faces. Examples of some north symbols are shown. The arrowhead points to the sun (north). Draw the symbol to a suitable size. Use a circle template for circular shapes.

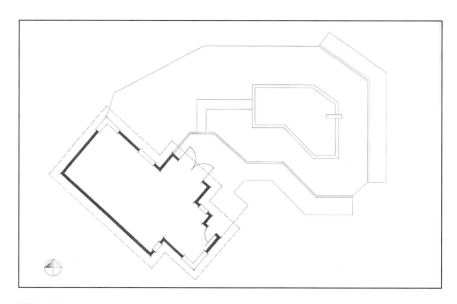

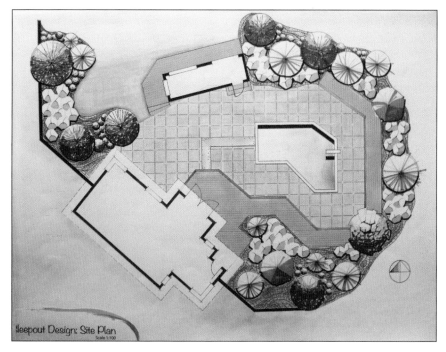

James Morrison, Year 10
Rendering media: *ProMarkers, colouring pencil, black fineline pen, eraser and pastel pencil*

Chelsea Evans, Year 13
Rendering media: *ProMarkers, colouring pencil, black fineline pen, eraser and chalk pastel*

ISBN: 9780170355575

2D landscape graphics

Site plans can be enhanced with careful drawing and rendering of landscape features such as trees, shrubs, pathways, etc. Examples of plan views of tree and shrub shapes and their colours are shown on these pages.

Media: *ProMarkers, Copic Markers, Kurecolor Markers, black fineline pen*

TIPS

- *Use a circle template* to lay out the planting scheme as a series of circles first, then sketch the plant forms inside the circles. Choose plant shapes that are visually interesting rather than looking real.
- Use a mix of colours such as browns, blues, reds and yellows, not only greens. Colours will add visual interest and bring the landscape to life.
- To give a 3D effect, place a shadow with a marker pen, around plantings and use darker tones opposite the light.
- Make large trees overlap smaller plantings and other features in the landscape.

Sandstone, Cinnamon, Cool Grey

Sandstone, Forest Green

Sandstone, Cinnamon, Light Grey

Oatmeal, Meadow Green, Cocoa, Light Grey

Sandstone, Meadow Green, Cool Grey

Pale Yellow, Shadow Mauve, Black

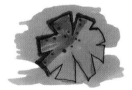

Grass Green, Sandstone

Forest Green, Sandstone, Black

Sandstone, Cocoa, Black

Matthew Beneka, Year 13

Rendering media:

Kurecolor Markers and ProMarkers
Colouring pencil
Black fineline pen
Chalk pastel

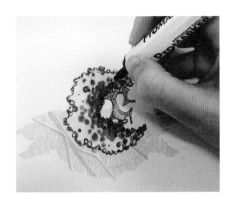

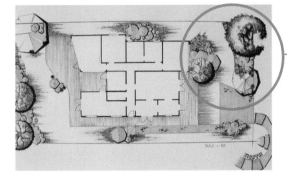

Enlarged detail shown below.

Hint: A circle template has been used to lay out the planting before drawing in the shapes and rendering.

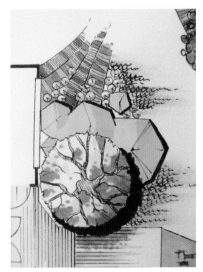

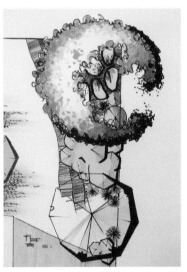

3D landscape graphics

Careful drawing and rendering of 3D pictorial drawings of trees and shrubs will provide visual interest and enhance the look of your work. Some examples of tree and foliage shapes and their colours, drawn by students, are shown.

Media: *ProMarkers, Copic Markers, black fineline pen, colouring pencils, watercolour paints*

TIPS

- Tree trunks should be *thin*, not thick — except where the tree is in the foreground and it is large, and you want to use it to frame a drawing.
- Make sure the size of a tree is in correct proportion to the drawing to give the correct sense of scale.
- Disguise where trunks enter the ground with a jagged line on a horizontal axis.
- When drawing rocks, use straight lines to define the shape.
- Use a grey marker to render the surfaces furthest from the light source.
- When drawing tree foliage, strive for an interesting shape to the canopy, not realism. A winter scene will avoid the need for drawing a canopy.
- When rendering tree foliage and shrubs, etc., make the bottom of areas, and those furthest away from the light, darker. Use the end of a paintbrush or marker pen to dab on a series of light and dark colours around the branch structure. Develop the interplay between light and shadow as opposed to trying to draw realistic leaves.
- Don't think in terms of only using green. Use a mix of colours such as browns, and yellows and blues, etc. Vary their values (light and dark).

Careful placement of large specimen trees, as shown in the perspective drawings below, heightens the sense of scale.

A large trunk, or part of a trunk, placed in the foreground is a good way of framing the drawing, much as a photographer would frame a photograph.

Thanks to the following students: Dustin Parkinson, Daniel Noyce, Troy Kent, Ashley Boyd, Chelsea Evans, Roxanne Bottom, Matthew Beneka and Gloria Li.

Elevations

Elevations are 2D drawings of a building. They show a view that looks directly at the side of a building and are labelled according to their direction to north.

They can be placed in any order on a page and are drawn to scale. Landscape features can be placed about each view, which may also be colour rendered. Some elevations and their rendering media are shown.

Rendering media: *ProMarkers, 4B pencil and eraser, coloured pencil, black fineline marker, pastel pencil*

Francis Ferguson, Year 13

Rendering media: *ProMarkers, fineline black marker, fineline white pen, white pastel pencil*

Rendering tips and techniques

Windows — Technique 1

STEP 1
Use a black fineline pen against an ellipse template to draw a curve through the glass parts.

STEP 2
Draw another curve beneath but on a different angle.

STEP 3
Use a black marker pen to render the top part only of the curved part.

STEP 4
Use a black fineline pen to shade in the depth of the window frame on the side opposite the light.

Windows — Technique 2

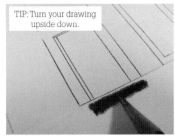

TIP: Turn your drawing upside down.

STEP 1
Smudge 4B pencil onto the edge of a piece of paper against the top of the window.

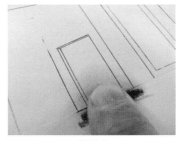

STEP 2
Use your finger to smudge the 4b into the top half only of the window. Use a circular motion.

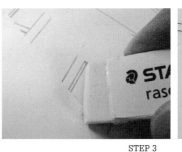

STEP 3
Erase the bottom half of the smudged area to make a sharp edge. Use an erasing shield or a paper edge.

Sky

TIP: Turn your drawing upside down.

STEP 1
Tear a piece of paper to make a jagged edge. Apply blue pastel pencil to the torn edge.

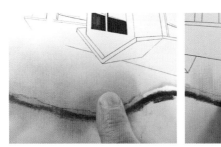

STEP 2
Smudge the pastel with your finger tip into the drawing using a circular motion. Vertical strokes will give a rainy-day look.

For an alternative look, smudge pastel from the roof line upwards to fade into white. Make clouds by scribbling and scrubbing with an eraser.

ISBN: 9780170355575

Concrete or plaster wall

STEP 1
Apply pastel pencil to the edge of a piece of paper.

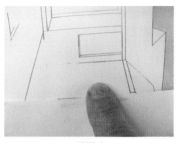

STEP 2
Smudge the pastel into the surface using circular strokes of your finger tip.

STEP 3
To provide tone, use the same technique to smudge 4B pencil over the walls opposite the light.

Grass

Use colouring pencil. Shade darker underneath foliage and next to features. Make your pencil strokes horizontal, not curved. Fade to white.

Tiles

STEP 1
Use colouring pencil. Smudge 4B pencil from a paper edge to make areas darker next to features.

STEP 2
Make reflections using an eraser and erasing shield to produce a wide and narrow vertical stripe.

STEP 3
Draw the tiles with a 0.05 black fineline pen. Make a white edge with an eraser and erasing shield.

Roof overhang shadow on walls

Apply 4B pencil to a paper edge then smudge it onto the walls. The paper edge should be parallel to the angle of the slope of the roof.

Completed rendered drawing

Rendering media: *colouring pencil, pastel pencil, 4B pencil, marker pens, fineline black pen*

Human figure

People are important in spatial designs. They give a sense of scale to the drawing and show the use or activity of the space.

Three things to consider when drawing human figures are **size**, **proportion** and **activity**.

Practise with stick figures of the correct proportions as shown below. Most people are approximately eight heads high.

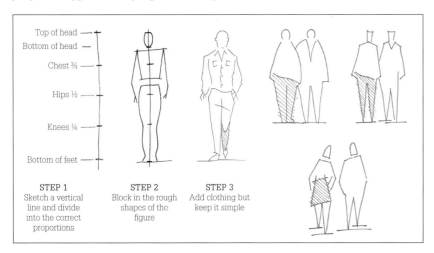

Top of head
Bottom of head
Chest ¾
Hips ½
Knees ¼
Bottom of feet

STEP 1
Sketch a vertical line and divide into the correct proportions

STEP 2
Block in the rough shapes of the figure

STEP 3
Add clothing but keep it simple

- Practise drawing people until you get the right proportions.
- Stylised people (above right) are simplified outlines of the human shape. They are quick and easy to draw and work well in either design sketches or presentation drawings. Shading on these sketches can indicate that the person is facing away from you.
- To achieve a sense of scale, it is a good idea to place eyes on the horizon line (shown right).
- Render the figure with shading on the sides opposite the light source.
- Don't add too many details to your people. Try to create a flowing line and give your people a gesture or movement.

Horizon line

SPATIAL DESIGN

Do this exercise to develop the following spatial design drawing skills:

- How to draw bubble diagrams.
- How to draw and render landscape and architectural graphics.
- How to draw site plans and floor plans of buildings.
- How to draw to scale, elevations and sectioned views of buildings.
- How to draw in one-point (parallel) perspective and two-point (angular) perspective.
- How to draw a rendered presentation drawing.

Note: To prepare students for a design assignment, this practice brief must be led by the teacher to show the class the layout and skills required for each page. Students may add their own design ideas. The work should not be rushed. Time: approximately six weeks.

EXERCISE: CONTAINER HOME SLEEPOUT

SITUATION
Shipping containers are used all over the world to create sustainable and effective architectural designs. Their affordability, mobility, durability and versatility make them a popular choice when in need of a small building.

TASK
Design a sleepout for yourself and your friends, from a shipping container to be located on the site shown on these pages, or one of your own choosing.

SPECIFICATIONS
The sleepout must have:
- easy access
- a deck that enables indoor-outdoor flow
- a sleeping area, bathroom, work desk and recreation area
- natural light (e.g. windows, sunroof, solar panels, etc.).

Work on your own A3 paper to produce the following set of drawings:

PAGE 1: A planning page — research, bubble diagram, line drawing.
PAGE 2: Ideation sketches of the sleepout.
PAGE 3: A site plan to show your sleepout and landscape graphics.
PAGE 4: Elevations and floor plan of the sleepout.
PAGE 5: A perspective presentation drawing.

Make title blocks for all the sheets. *These may be computer generated or done by hand.*

Present the drawings stapled together, down the left side, landscape format, with a front cover that you have made. *This may be computer generated.*

PAGE 1 PLANNING

1 Paste four or more images (research) of existing container homes onto the page. Write some notes that discuss the aesthetics and function of each.

2 Produce a bubble diagram to show spatial relationships, both indoor and outdoor.

3 From your bubble diagram, produce a line diagram of your sleepout idea.

Notes
- It is a good idea to list the specifications so that you don't forget them when planning.
- Use a circle or ellipse template to draw the bubbles.
- The line diagram may be drawn with a straight edge.
- Notes should be placed on a horizontal guide line.
- Outline the drawings with a black fineline pen.
- Choose a colour palette and apply this to the page.

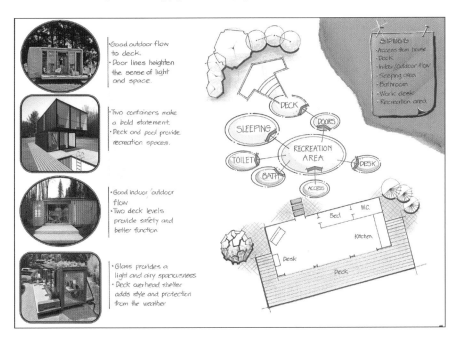

A border and title block can be made on a computer. The page can then be printed ready for the drawings. Research has been found on the internet and cut and pasted onto the page.

Rendering media: *black fineline pen for outlines, colouring pencils, marker pens, pastel pencils*

PAGE 2 IDEATION

1 Produce a series of three perspective sketches of your ideas for the designs of your sleepout. Show the following:
- Doors and windows.
- Deck and any associated features (all-weather cover, built-in seating, etc.).

2 Paste research about the sketches and use brief notes that explain any thinking not clear visually.

3 Outline the drawings with a fineline black pen and show some quick rendering.

Notes
- A minimum of *three* sketches should be shown. Place them within a background.
- The sketches may be constructed with a straight edge then drawn over the top freehand.
- The sketches must consider *proportion* and *viewpoint* and show crating clearly.
- Notes and any freehand title block should be placed between guide lines.
- Keep notes horizontal and use lower-case lettering.
- Indicate your chosen design and show a human figure.

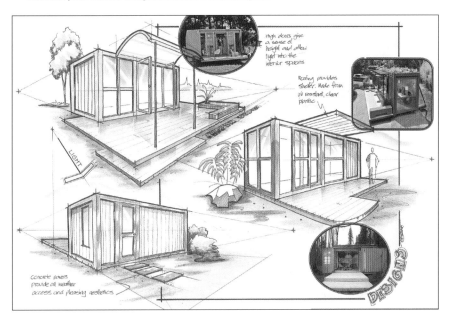

Note the handwritten title, the use of thick and thin lines, a human figure, and the attention given to viewpoint (position of the horizon lines).

Rendering media: *black fineline pen for outlines, colouring pencil, 4B pencil for shadows, marker pens, pastel pencil (for the sky)*

PAGE 3 SITE PLAN

1 Using a scale of 1:100, either copy and draw on an A3 page the site plan shown below, or make up one of your own, to show the floor plan of your sleepout from Page 1 (Planning).

2 Outline the drawing with a fineline black pen.

3 Design and draw landscape graphics on the site plan. Show the north symbol and colour render the drawing.

Notes
- The site is flat.
- The sleepout is secured into the ground on four concrete posts — see Page 4 (Elevations and floor plan).
- The sleepout should be sited so that the main entry is easily accessible and provides flow to outdoor areas.
- Container dimensions: L 7.500 m, W 2.500 m.

Here is an example site plan with a border and handwritten title.

The plants have been drawn as circles first, using a circle template. The shapes have then been drawn over the circles with a fineline black pen in preparation for colour rendering.

A rendered example of the site and the media used is shown on the opposite page.

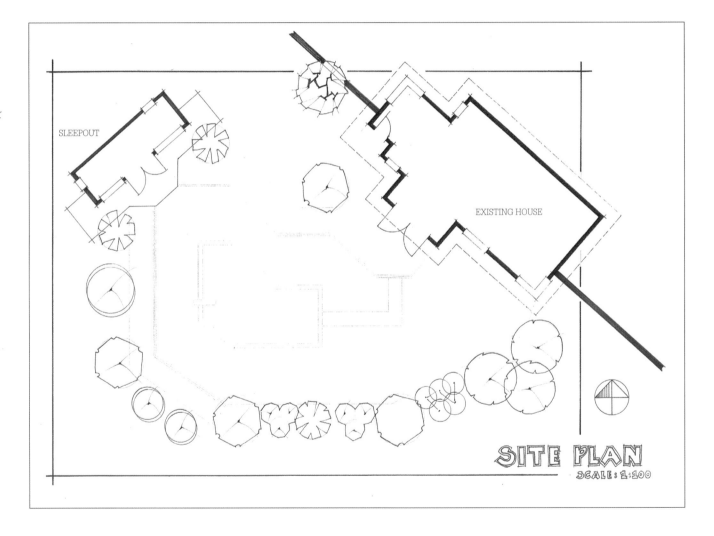

SLEEPOUT

EXISTING HOUSE

SITE PLAN
SCALE: 1:100

ISBN: 9780170355575

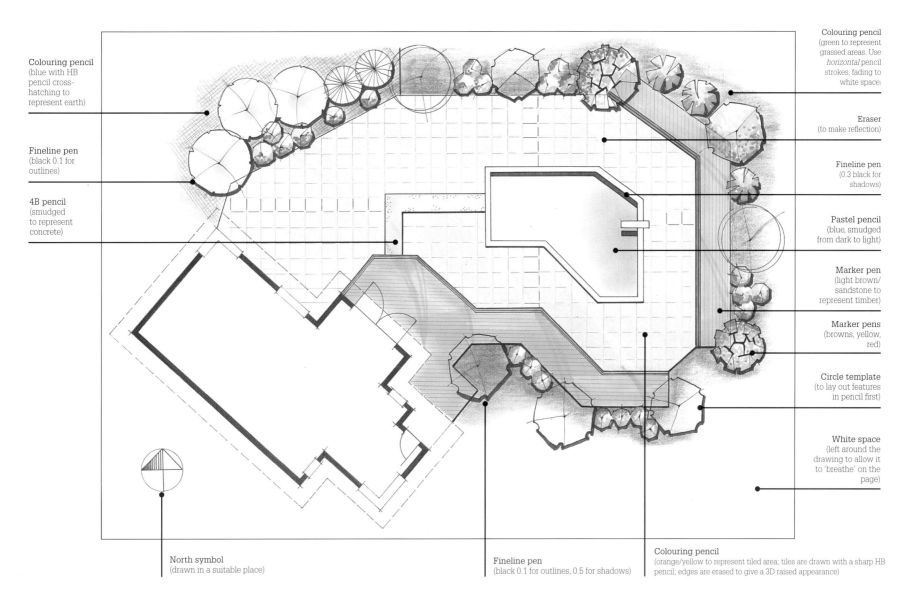

Colouring pencil
(blue with HB pencil cross-hatching to represent earth)

Fineline pen
(black 0.1 for outlines)

4B pencil
(smudged to represent concrete)

Colouring pencil
(green to represent grassed areas. Use *horizontal* pencil strokes, fading to white space)

Eraser
(to make reflection)

Fineline pen
(0.3 black for shadows)

Pastel pencil
(blue, smudged from dark to light)

Marker pen
(light brown/sandstone to represent timber)

Marker pens
(browns, yellow, red)

Circle template
(to lay out features in pencil first)

White space
(left around the drawing to allow it to 'breathe' on the page)

North symbol
(drawn in a suitable place)

Fineline pen
(black 0.1 for outlines, 0.5 for shadows)

Colouring pencil
(orange/yellow to represent tiled area; tiles are drawn with a sharp HB pencil; edges are erased to give a 3D raised appearance)

Notes

- The sleepout on this drawing has not been shown. You should draw this in your chosen place. (See previous page.)
- Plan your landscape graphics for visual appeal. Use a circle template to lay out plants and trees, in pencil first.
- Use a mixture of large and small plantings, grouped in a pleasing manner.
- Large trees should overlap smaller plantings. Place a shadow beneath plants, opposite the light source (north).

PAGE 4 ELEVATIONS AND FLOOR PLAN

1. To a scale of **1:50**, using instruments, draw two elevations of the sleepout. *Use the layout shown.*
2. To a scale of **1:25**, using instruments, draw the floor plan of the sleepout.
3. Show three dimensions and the following symbols on the floor plan: *a one-way and a two-way switch, lights, socket outlets, distribution board and underground point of entry, cold water tap, doors and windows, interior features — bed, sink, furniture, toilet, etc.*

Container specifications:

- Overall dimensions are 7.500L x 2.500W x 3.000H.
- Rests on four 500 mm square concrete feet, raised 100 above the ground line.
- Ground clearance: 300 mm.
- Construction consists of 200 x 100 steel frames with timber sheet walls.
- Container walls as seen in the floor plan are 100 mm thick.

Notes

- The views shown are based on the design seen in the presentation drawing opposite.
- The deck has not been shown.
- Arrows indicate set-out distance from edge of paper.

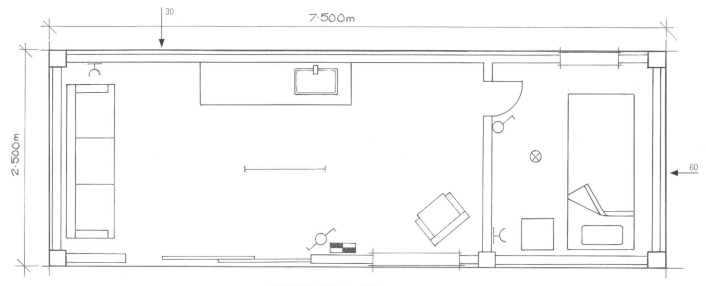

FLOOR PLAN Scale 1:25

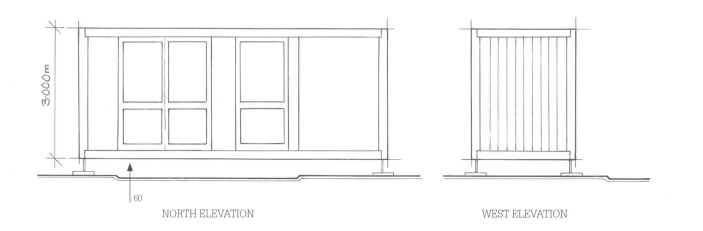

NORTH ELEVATION

WEST ELEVATION

ELEVATIONS Scale 1:50

ISBN: 9780170355575

PAGE 5 PERSPECTIVE PRESENTATION DRAWING

1 From the set-out measurements given, and following the numbered steps 1–3, use instruments to draw a well-proportioned, two-point perspective view of the sleepout.

2 Show the deck, some of the surrounding landscape graphics you designed on the site plan, and a human figure.

3 Render the drawing for visual impact. Suggested media: *colouring pencils, marker pens, pastel pencils, black fineline markers 0.1 and 03, eraser and erasing shield, 4B pencil, white highlight pen*

Horizon line and vanishing points set-out measurements are from the edge of an A3 sheet.

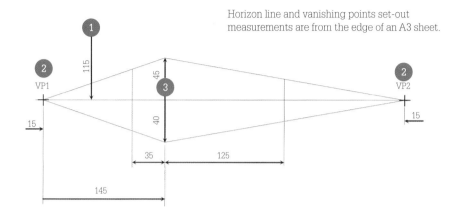

SITUATION

Holidays, visits to friends, family and whanau, or staying on a marae, etc., all require luggage to contain personal belongings.

BRIEF: Design a suitcase for your trip or stay.
DURATION: Six weeks (class time)

REQUIREMENTS

1   Design research. *Cut and pasted among your sketches.*
2   Ideation of the suitcase. *Your sketches should show five stages: thumbnails, exploratives, alternatives, thinking, and technical sketches. Explore two alternatives that inform your design. Consider human factors. Indicate your chosen design. Two or three A3 pages.*
3   Place notes where necessary, about your research and sketches, to explain your thinking and details that are not clear visually. Discuss *aesthetics* and *function*.

4   Draw with instruments a detailed, scaled orthographic projection of your suitcase design. Show the following:
   - A plan, one end elevation and a front elevation. *(Section one of the views to show technical detail, by use of a cutting plane.)*
   - Hidden detail where necessary to clarify construction.
   - Six main dimensions.
   - A title block with the scale and projection symbol.
   - The reference line and labelled views.
   - A parts list that shows materials and numbered components about the views.
5   Produce on a computer or by hand a rendered presentation drawing of your design. *(A straight edge may be used to aid in the construction of the drawing.)*

DESIGN SPECIFICATIONS

The suitcase must:
- Suit a variety of activities during your trip and transform in some way, in either shape or number of components.
- Consider aesthetics and function.
   Function: durability, fitness for purpose, construction, user friendliness, safety, ergonomics, materials.
   Aesthetics: style, proportion, form, finish, texture.

Ideation

THUMBNAILS AND EXPLORATIVE SKETCHES

Product design is the hardest area to write a brief for. Not only because of the depth of content required, but to ensure gender neutrality. The set of drawings featured demonstrate how a student can take a simple idea and transform it into something exciting and individual.

The brief should be introduced to students in stages. Each key stage must be preceded by skill-building exercises, particularly to develop more sophisticated sketching techniques.

Careful attention needs to be given to ensuring a mix of 2D and 3D sketches, pleasing page layout, and a range of viewpoints. Lots of white space should be left to allow sketches to 'breathe'.

Sketches have been done in pencil first, then inked in with fineline black pens. A colour palette has been chosen for each page. Rendering is intentionally minimalist, reflecting the skills of the students yet still providing information about shape and surface texture of the designs.

Use of black ballpoint pen enhances thumbnails. Effective use of craft paper gives the pages a 'work in progress' feel. The use of craft paper, or any paper other than white, is 'non scary', and should be encouraged. This allows the student to develop confidence in their sketching, and focus on the idea, instead of worrying whether or not the sketch is good enough or whether they are going to make a mistake. (There are no mistakes in ideation sketching.)

Font styles for handwritten titles needs to be taught and practised beforehand.

SKILLS NEEDED TO PRECEDE THIS PAGE:

Planning sheet.

Freehand design sketching: thumbnails use a black ballpoint pen on craft paper, exploratives use a mix of 2D and 3D, varying viewpoint, backgrounds.

Crating (proportions) and line hierarchy.

Rendering techniques — light source, tonal changes, shape and surface texture.

Colour palette considered.

Research and simple design notes that explain thinking not clear visually.

Media technique: marker pens, black fineline pens, pastel pencils.

Links to: pages 75–81, 88–91

ISBN: 9780170355575

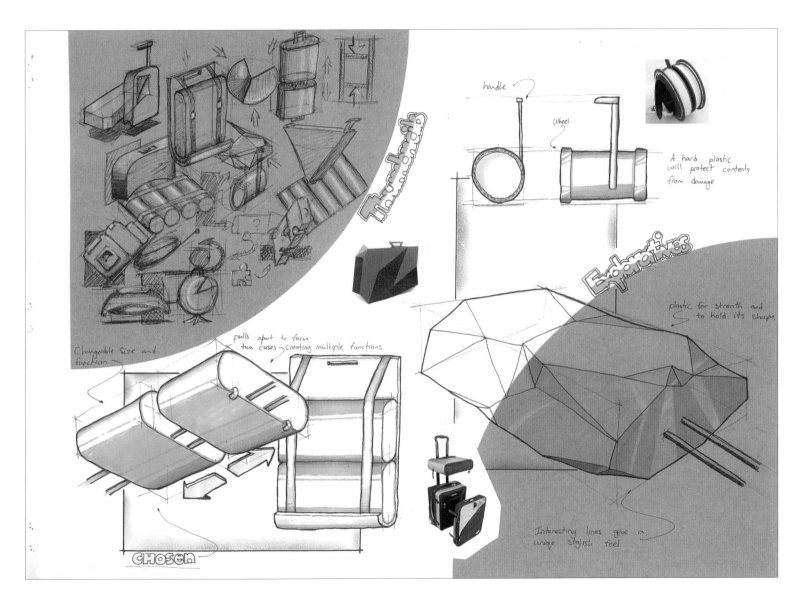

handle

wheel

A hard plastic will protect contents from damage

Thumbnails

Explorations

plastic for strenth and to hold its shape.

Changeable size and function

pulls apart to form two cases - creating multiple functions

Interesting lines give a unique stylish feel

CHOSEN

Before I started this page I sketched out a rough layout first. I knew my design was quite complicated, so this helped guide my eye around the page to show my idea development.

Because I had a good idea of what I wanted my design to look like, I didn't do as many thumbnails as I should. If I was to do the page again, I would have added more.

Even though I had done design sketches before, I needed to revise how to sketch in 3D and 2D from different viewpoints.

To make my design stand out I used simple, subtle rendering. I was a bit worried at first that if I used too much colour it would have overpowered the page. I used a blue and grey colour theme to represent the materials the design could be made from.

Sometimes I found it difficult to know when enough was enough.

Taylor Plank

The thing I found hard about my first page was where to start. So I did a practice layout page first. From this I got the idea of dividing the page into three parts on an angle.

I first of all glued brown paper to the top and bottom parts on an angle. The middle part of white paper was where I wanted to show my main design ideas.

My views are drawn in perspective, isometric and 2D orthographic. The hardest part was getting the right proportions.

I made steeper angles on the perspective view to show the top more.

I liked using the brown paper for my thumbnails because of the effects you can get with a ballpoint pen.

For the rendering I used marker pens.

If I was to do the page again I would have chosen better colours. I wished I had thought more about this before I started.

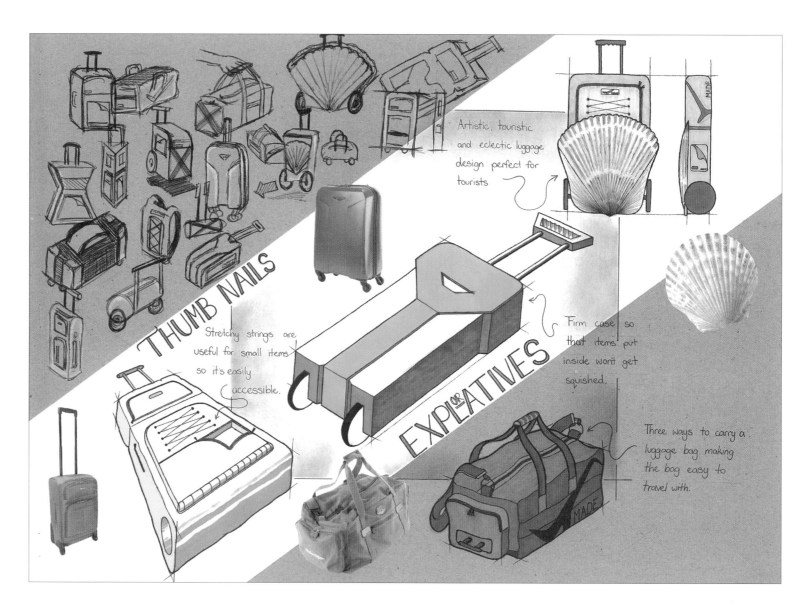

THUMB NAILS

Artistic, touristic and eclectic luggage design perfect for tourists

Stretchy strings are useful for small items so it's easily accessible.

Firm case so that items put inside won't get squished.

EXPLATIVES OR

Three ways to carry a luggage bag making the bag easy to travel with.

ISBN: 9780170355575   PHOTOCOPYING OF THIS PAGE IS RESTRICTED UNDER LAW.

Handle clips back into suitcase for good compactability

Handles provide an alternate way to carry bag

THUMBNAILS

EXPLORATIONS

Once I got started, this page wasn't as hard as I thought it would be.

I experimented with different viewpoints first before doing the sketches to get the best views.

The hardest part was making the crates the right sizes, especially for the designs with the handles.

I added shadows underneath and rendered all the sketches using marker pens. I had to do some practice first and used mainly blues and grey colours.

I like the way the page turned out, especially the arrows to show how the handles work.

If I was to do the page again I would leave more room to write the titles, but they turned out okay in the end.

SITUATION

Holidays, visits to friends, family and whanau, or staying on a marae, etc., all require luggage to contain personal belongings.

BRIEF: Design a suitcase for your trip or stay.
DURATION: Six weeks (class time)

REQUIREMENTS

1   Design research. *Cut and pasted among your sketches.*
2   Ideation of the suitcase. *Your sketches should show five stages: thumbnails, exploratives, alternatives, thinking, and technical sketches. Explore two alternatives that inform your design. Consider human factors. Indicate your chosen design. Two or three A3 pages.*
3   Place notes where necessary, about your research and sketches, to explain your thinking and details that are not clear visually. Discuss *aesthetics* and *function*.

4   Draw with instruments a detailed, scaled orthographic projection of your suitcase design. Show the following:
 ▪ A plan, one end elevation and a front elevation. *(Section one of the views to show technical detail, by use of a cutting plane.)*
 ▪ Hidden detail where necessary to clarify construction.
 ▪ Six main dimensions.
 ▪ A title block with the scale and projection symbol.
 ▪ The reference line and labelled views.
 ▪ A parts list that shows materials and numbered components about the views.
5   Produce on a computer or by hand a rendered presentation drawing of your design. *(A straight edge may be used to aid in the construction of the drawing.)*

DESIGN SPECIFICATIONS

The suitcase must:
 ▪ Suit a variety of activities during your trip and transform in some way, in either shape or number of components.
 ▪ Consider aesthetics and function.
   Function: durability, fitness for purpose, construction, user friendliness, safety, ergonomics, materials.
   Aesthetics: style, proportion, form, finish, texture.

Ideation (cont.)

ALTERNATIVES AND THINKING SKETCHES

Through the exploration of alternative ideas, by researching objects that have nothing to do with the design, allows students to extend their thinking to better inform the design. A lesson by the teacher in bio-mimickry promotes thinking in the right direction.

Note the continuation of the same sketching style, using 2D and 3D sketches, and the dramatic use of viewpoint to highlight shape and form. Crating is clearly defined, offering correct proportion, while the addition of shadows and backgrounds once again anchor the drawings to the page. Note arrows that show movement and/or function.

Although not necessary, writing down the specifications is a good idea to ensure no requirements are missed out.

3D exploded and 2D sectioned views convey important information about construction of the design (technical sketches). The all-important inclusion of human factors and dimensions determine fitness for purpose.

Once again, minimalist marker rendering conveys tone, shape and surface texture effectively. Good page layout, effective titles, and generous white space allow sketches to 'breathe' on the page.

SKILLS NEEDED TO PRECEDE THIS PAGE:

Planning sheet used.
Freehand design sketching in 2D and 3D with clearly seen crating (proportions) and line hierarchy.
Viewpoints are considered for visual interest and to better show the ideas.
Arrows indicate movement and/or function.
Research is ongoing to inform ideas.
Design notes explain thinking not clear visually.
Rendering techniques — minimalist, indicating the light source, tonal changes, form and surface texture.
Colour palette — soft tones.
Links to: pages 75–81, 88–91

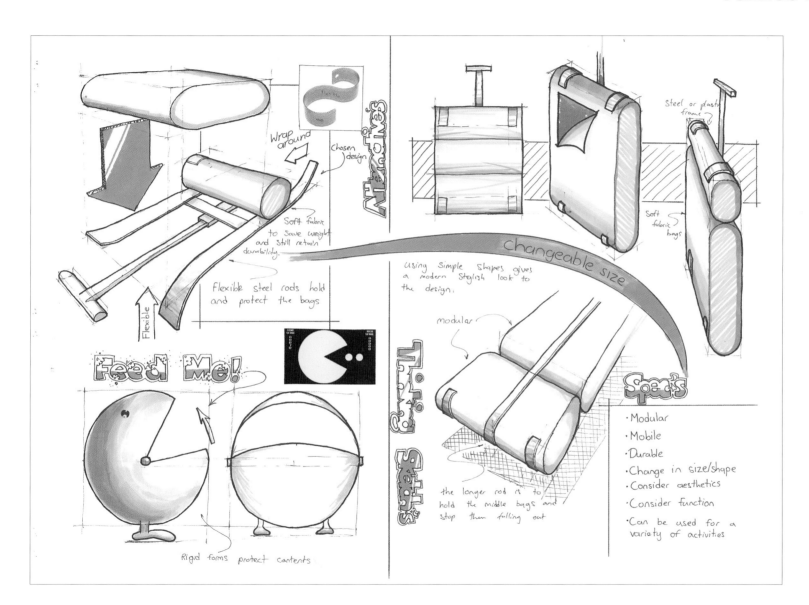

For this page I looked at some alternative research. I found some good ideas on the internet in the form of the flexible metal band and the Pac-Man logo from old video games.

I looked at both of these to see how they could be used in the design. I liked the flexible metal strap idea best and used this in the final design.

Once I had my research I did a page layout again to make sure I could fit all the drawings I wanted to show onto the page.

To make the drawings look more interesting I used steeper viewpoints on some of the 3D sketches and used arrows to show how parts fit together.

I wrote down the specs so that I didn't forget any. I found this helpful.

I made the page in two parts and put some backgrounds behind the sketches to show a flow of ideas.

Taylor Plank

For this page I looked at bark as alternative research. I chose this because I like the texture of it and also wanted my design to embrace the environment.

I also used bark as the colour for my design and for the colour palette of my page.

Even though I planned the page layout first, I still ran out of room. So I used a lift-up flap in the top right corner to show the sizes of my design.

The exploded drawing under the flap (shown below) shows the detachable part of the design.

I had to practise drawing people and hands to show human factors.

I used brown paper again to match the look of the first page. It is also the colour of bark.

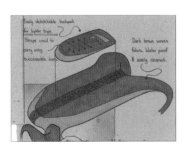

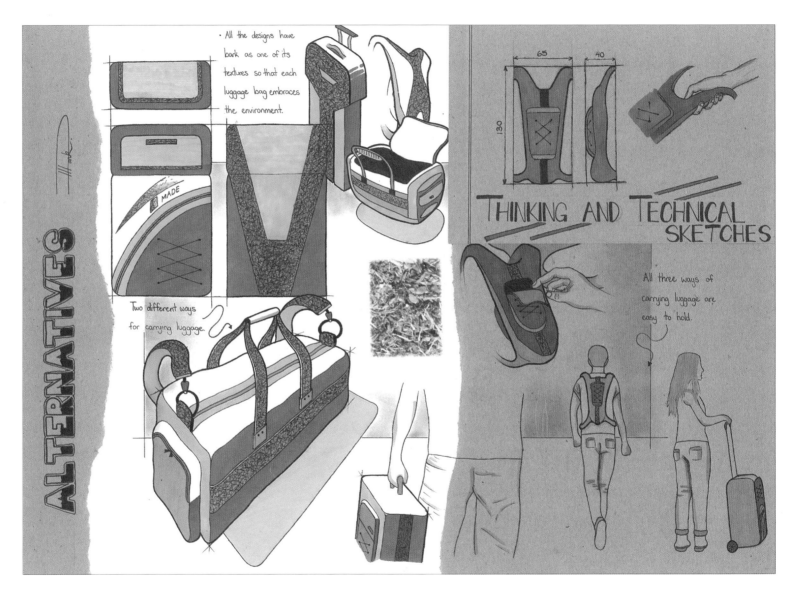

Note that technical sketches have been included on this page (see next section).

ISBN: 9780170355575   PHOTOCOPYING OF THIS PAGE IS RESTRICTED UNDER LAW.

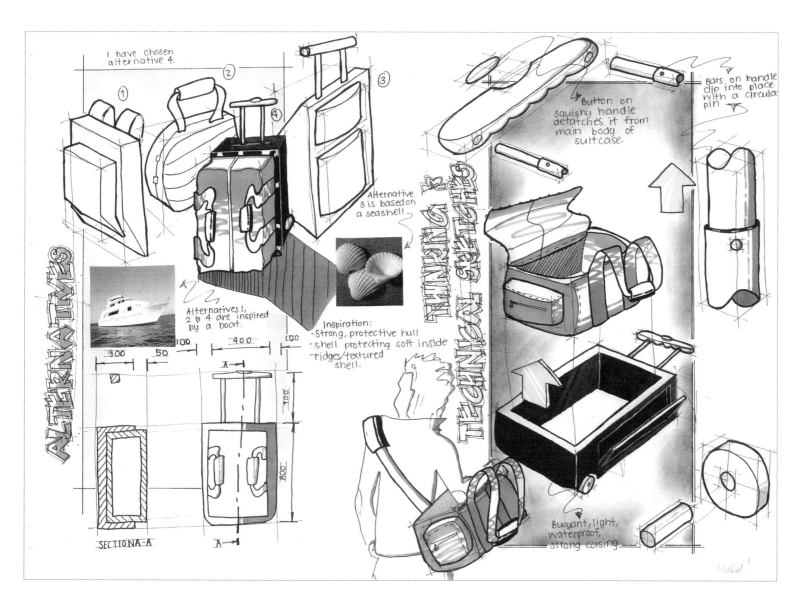

I have chosen alternative 4.

① ② ③ ④

ALTERNATIVES

Alternatives 1, 2 & 4 are inspired by a boat.

Alternative 3 is based on a seashell.

Inspiration:
- Strong, protective hull
- shell protecting soft inside
- ridges/textured shell.

300 50 100 400 100

100

300

SECTION A-A

A   A

THINKING & TECHNICAL SKETCHES

Button on squishy handle detaches it from main body of suitcase.

Bars on handle clip into place with a circular pin

Buoyant, light, waterproof, strong casing

The first thing I did was do a page layout. I had a lot of details I wanted to show and needed to make sure it could all fit on the page.

Because my luggage is used for holidays, I used boat and shell research to help me decide on the final design.

I made the 3D sketches overlap each other. I saw this in some books in my classroom. I like the way you can show your ideas changing more clearly.

If I was to do the page again, I would put the technical sketches on a separate page. I ran out of room so made a lift-up flap (seen below) to show human factors.

Rendering the curved parts and fabric was hard, but I kept it simple using marker pens. Putting a background behind each set of sketches made the page easier to read.

SITUATION

Holidays, visits to friends, family and whanau, or staying on a marae, etc., all require luggage to contain personal belongings.

BRIEF: Design a suitcase for your trip or stay.
DURATION: Six weeks (class time)

REQUIREMENTS

1   Design research. *Cut and pasted among your sketches.*
2   Ideation of the suitcase. *Your sketches should show five stages: thumbnails, exploratives, alternatives, thinking, and technical sketches. Explore two alternatives that inform your design. Consider human factors. Indicate your chosen design. Two or three A3 pages.*
3   Place notes where necessary, about your research and sketches, to explain your thinking and details that are not clear visually. Discuss *aesthetics* and *function*.

4   Draw with instruments a detailed, scaled orthographic projection of your suitcase design. Show the following:
   ▪   A plan, one end elevation and a front elevation. *(Section one of the views to show technical detail, by use of a cutting plane.)*
   ▪   Hidden detail where necessary to clarify construction.
   ▪   Six main dimensions.
   ▪   A title block with the scale and projection symbol.
   ▪   The reference line and labelled views.
   ▪   A parts list that shows materials and numbered components about the views.
5   Produce on a computer or by hand a rendered presentation drawing of your design. *(A straight edge may be used to aid in the construction of the drawing.)*

DESIGN SPECIFICATIONS

The suitcase must:
▪   Suit a variety of activities during your trip and transform in some way, in either shape or number of components.
▪   Consider aesthetics and function.
   Function: durability, fitness for purpose, construction, user friendliness, safety, ergonomics, materials.
   Aesthetics: style, proportion, form, finish, texture.

Ideation (cont.)

TECHNICAL SKETCHES

Now is the time to think about construction details (technical detail) of how parts of the design could be made.

Technical sketches need to be a mix of 3D exploded isometric and 2D sectioned views of parts.

James has shown this extremely well, confirming his understanding not only of how his design could be made, but also his advanced knowledge of sketching technique.

He has used an entire A3 page to allow the size of each sketch to be large enough to show detail clearly, and allow for correct sketch construction methods. This is an important consideration, and particularly effective in his 2D sectioned drawings.

Note also the inclusion of human factors to determine fitness for purpose.

Good use of line hierarchy (thick and thin lines), minimalist rendering, colour palette and titles, continue the theme from earlier pages.

A good idea, if more room is required for ideas, is to introduce a lift-up flap of craft paper as shown opposite.

SKILLS NEEDED TO PRECEDE THIS PAGE:

Planning sheet used.

Freehand design sketching in 2D and 3D and line hierarchy.

Viewpoints are considered for visual interest and to better show the ideas.

Correct 'exploded' techniques. The parts are pulled apart in the direction in which they would be assembled.

Correct 2D cross-hatching technique to show internal detail.

Simple notes explain thinking not clear visually.

Rendering techniques — minimalist, indicating the light source, tonal changes, form and surface texture.

Colour palette — soft tones.

Links to: pages 75–81

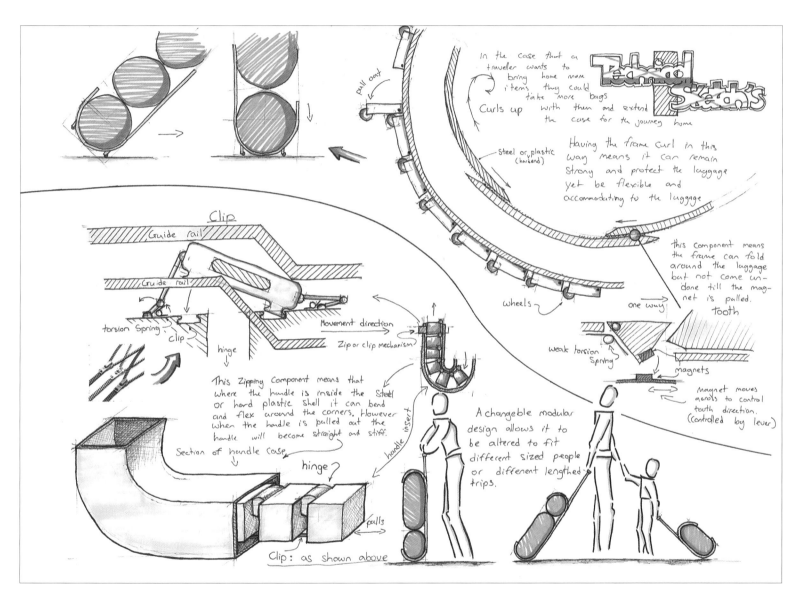

In the case that a traveller wants to bring home more items they could take more bags with them and extend the case for the journey home

Curls up

Technical Sketches

pull out

Steel or plastic (hardened)

Having the frame curl in this way means it can remain strong and protect the luggage yet be flexible and accommodating to the luggage

this component means the frame can fold around the luggage but not come undone till the magnet is pulled.

one way

Tooth

wheels

Clip

Guide rail

Guide rail

torsion spring

Clip

hinge

Movement direction

Zip or clip mechanism

weak torsion spring

magnets

magnet moves across to control tooth direction. (controlled by lever)

This Zipping Component means that where the handle is inside the steel or hard plastic shell it can bend and flex around the corners. However when the handle is pulled out the handle will become straight and stiff.

handle insert

A changeable modular design allows it to be altered to fit different sized people or different lengthed trips.

Section of handle case

hinge?

pulls

Clip: as shown above

For this page I had to research how my design would be made. This took a long time but I wanted to make sure that my design could actually work as well as be made.

I did a rough page layout first, to get all the information on the page looking right.

It was hard to show it in drawings so I used 2D section views to show it clearer.

I used different coloured paper to make important features stand out like on the black lift-up flap (see below). I used a white pen to draw with on this.

Because I forgot to draw an exploded view, to meet the brief requirements, I also placed it on the lift-up flap attached to the top of the page.

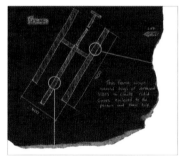

SITUATION

Holidays, visits to friends, family and whanau, or staying on a marae, etc., all require luggage to contain personal belongings.

BRIEF: Design a suitcase for your trip or stay.
DURATION: Six weeks (class time)

REQUIREMENTS

1   Design research. *Cut and pasted among your sketches.*
2   Ideation of the suitcase. *Your sketches should show five stages: thumbnails, exploratives, alternatives, thinking, and technical sketches. Explore two alternatives that inform your design. Consider human factors. Indicate your chosen design. Two or three A3 pages.*
3   Place notes where necessary, about your research and sketches, to explain your thinking and details that are not clear visually. Discuss *aesthetics* and *function.*

4   Draw with instruments a detailed, scaled orthographic projection of your suitcase design. Show the following:
   ▪   A plan, one end elevation and a front elevation. *(Section one of the views to show technical detail, by use of a cutting plane.)*
   ▪   Hidden detail where necessary to clarify construction.
   ▪   Six main dimensions.
   ▪   A title block with the scale and projection symbol.
   ▪   The reference line and labelled views.
   ▪   A parts list that shows materials and numbered components about the views.

5   Produce on a computer or by hand a rendered presentation drawing of your design. *(A straight edge may be used to aid in the construction of the drawing.)*

DESIGN SPECIFICATIONS

The suitcase must:
▪   Suit a variety of activities during your trip and transform in some way, in either shape or number of components.
▪   Consider aesthetics and function.
   Function: durability, fitness for purpose, construction, user friendliness, safety, ergonomics, materials.
   Aesthetics: style, proportion, form, finish, texture.

Instrumental drawing

ORTHOGRAPHIC PROJECTION

Do not overlook the importance, in product design, of producing a third angle orthographic projection of the final design. This drawing system provides important, accurate, scaled information on sizes and construction that could allow the design to be manufactured.

This drawing system also provides students with the full range of drawing skills, and complements the freedom of ideation pages.

Before these drawings can be attempted, comprehensive classwork exercises must be undertaken, directed at every step by the teacher. Orthographic projection needs to be introduced in earlier years of study. Students must not be expected to draw their designs in this format without any prior knowledge.

All drawings featured are excellent examples of the precision of line and printing standards required. They show a wide range of essential drawings skills: correct line weights and types, dimensioning, a cutting plane, notation, a title block, printing, cross-hatching and a parts list.

Three views are normally enough, but because of the complicated amount of detail in James's design, he has introduced an extra rear view elevation. Caroline has used portrait page orientation to better suit the dimensions of her design.

Knowledge of how to make a parts list must be taught. It can be produced on a computer, then cut and pasted onto the drawing.

⊚ SKILLS NEEDED TO PRECEDE THIS PAGE:

Use of a drawing board and instruments.
Line weights and line standards (2H pencil for all lines).
Title block, scales and projection symbol.
Printing standards (HB pencil), correct dimensioning techniques for third angle projection.
Cutting plane and cross-hatching, hidden detail, reference lines, notation.
A parts list with numbered components about the views.
Views labelled.
Links to: pages 58–64, 71

James Morrison

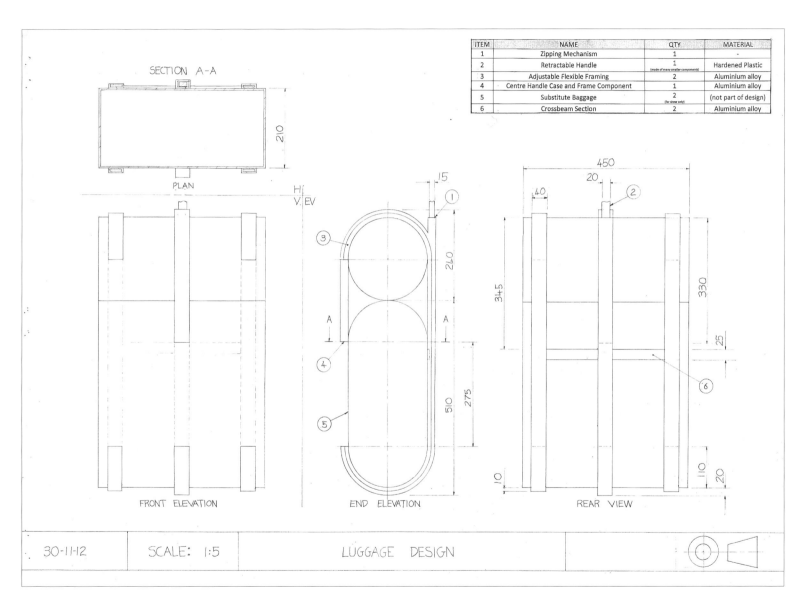

ITEM	NAME	QTY	MATERIAL
1	Zipping Mechanism	1	-
2	Retractable Handle	1 (made of many smaller components)	Hardened Plastic
3	Adjustable Flexible Framing	2	Aluminium alloy
4	Centre Handle Case and Frame Component	1	Aluminium alloy
5	Substitute Baggage	2 (for show only)	(not part of design)
6	Crossbeam Section	2	Aluminium alloy

SECTION A-A

210

PLAN

FRONT ELEVATION

END ELEVATION

REAR VIEW

H V EV

450
20
40
15
345
330
275
510
210
25
110
20
10

30-11-12   SCALE: 1:5   LUGGAGE DESIGN

This was a hard drawing. I had to spend lots of time getting the right scale and sizing and placing the orthographic views so that they filled the page.

I planned the page first by doing a practice sheet to get the layout of the views right.

I was glad I had done lots of orthographic drawings in class beforehand.

I had to cut off the handle to make it fit.

Because my design had details on the back I wanted to show, I added an extra rear view beside the right end elevation.

I had to be careful with my linework to make it the best I could. I used a clutch pencil with a 2H lead for all the lines, and an HB lead for the printing.

I made the parts list on a computer and pasted it on the page at the end.

Taylor Plank

I'm glad I had done orthographic drawings in class beforehand, and in Year 9. Although I knew how to do this drawing, the curved parts made it more complicated.

The first thing I did was do a practice page to get the right scale and layout. I wanted to make the drawing bigger, and also wanted to show two end elevations. This meant making the scale smaller, but it worked out in the end.

I got the sizes from measuring my school bag.

I drew all the boxes first, then used ordinates to plot the mirror image of the curved sides. I knew how to do this from other drawings I had done. Then I used a French curve to draw the curved sides.

The part that took the longest was getting all the lines and my printing neat and accurate. I used my clutch pencil and a 2H lead.

I made the parts list on a computer using Word, then pasted it onto the page.

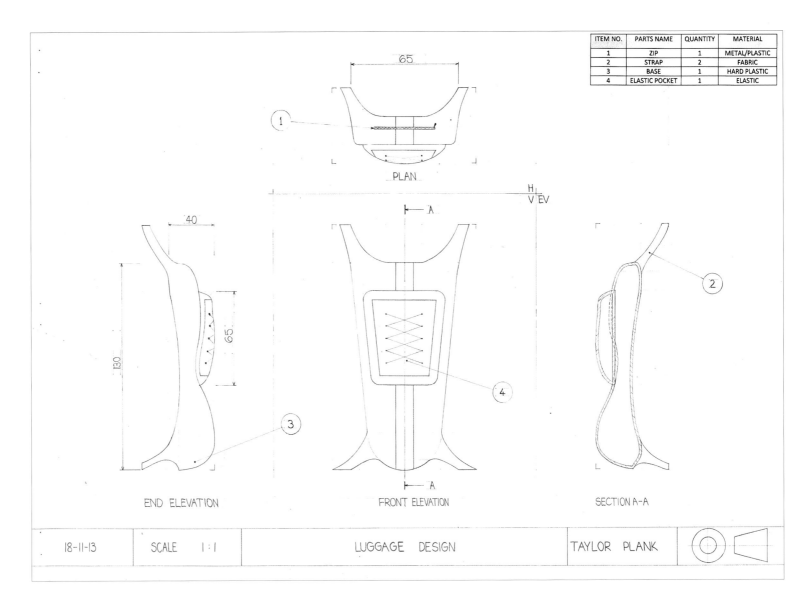

ITEM NO.	PARTS NAME	QUANTITY	MATERIAL
1	ZIP	1	METAL/PLASTIC
2	STRAP	2	FABRIC
3	BASE	1	HARD PLASTIC
4	ELASTIC POCKET	1	ELASTIC

PLAN

END ELEVATION

FRONT ELEVATION

SECTION A-A

18-11-13 | SCALE 1:1 | LUGGAGE DESIGN | TAYLOR PLANK

Caroline Webb

The hardest part about this drawing was getting it to fit the page.

I wanted to show two end elevations, but they wouldn't fit the landscape page format. So I turned the page around and did it in portrait.

This worked really well and I was able to make the scale bigger and show more detail in the sectioned view.

I had to cut the middle part from the handle to get it to fit though, and show a break symbol. This also showed that the handle was round, so it worked out really well.

I took a lot of time to get my lines right. The hard part was doing the cross-hatching lines. I used my 45° set square for these.

I bought a clutch pencil especially to do this drawing. I'm glad I did as it helped a lot.

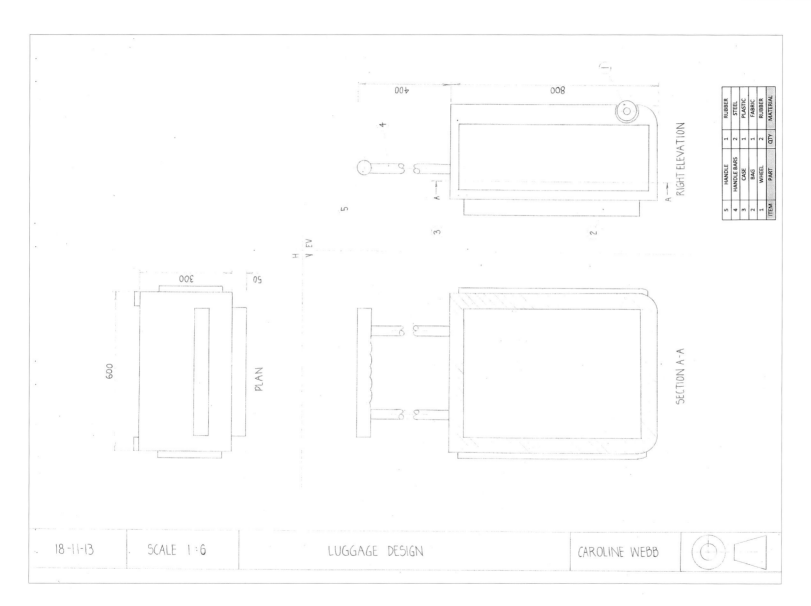

ITEM	PART	QTY	MATERIAL
5	HANDLE	1	RUBBER
4	HANDLE BARS	2	STEEL
3	CASE	1	PLASTIC
2	BAG	1	FABRIC
1	WHEEL	2	RUBBER

RIGHT ELEVATION

400 800

PLAN

300 50 600

SECTION A-A

18-11-13	SCALE 1:6	LUGGAGE DESIGN	CAROLINE WEBB

SITUATION

Holidays, visits to friends, family and whanau, or staying on a marae, etc., all require luggage to contain personal belongings.

BRIEF: Design a suitcase for your trip or stay.
DURATION: Six weeks (class time)

REQUIREMENTS

1   Design research. *Cut and pasted among your sketches.*
2   Ideation of the suitcase. *Your sketches should show five stages: thumbnails, exploratives, alternatives, thinking, and technical sketches. Explore two alternatives that inform your design. Consider human factors. Indicate your chosen design. Two or three A3 pages.*
3   Place notes where necessary, about your research and sketches, to explain your thinking and details that are not clear visually. Discuss *aesthetics* and *function*.
4   Draw with instruments a detailed, scaled orthographic projection of your suitcase design. Show the following:
    - A plan, one end elevation and a front elevation. *(Section one of the views to show technical detail, by use of a cutting plane.)*
    - Hidden detail where necessary to clarify construction.
    - Six main dimensions.
    - A title block with the scale and projection symbol.
    - The reference line and labelled views.
    - A parts list that shows materials and numbered components about the views.
5   Produce on a computer or by hand a rendered presentation drawing of your design. *(A straight edge may be used to aid in the construction of the drawing.)*

DESIGN SPECIFICATIONS

The suitcase must:
- Suit a variety of activities during your trip and transform in some way, in either shape or number of components.
- Consider aesthetics and function.
  Function: durability, fitness for purpose, construction, user friendliness, safety, ergonomics, materials.
  Aesthetics: style, proportion, form, finish, texture.

Presentation drawing

Time now to show off student talent and acquired skills, on a final presentation drawing that will 'sell' the design. This drawing needs to have the 'wow' factor. The most important consideration in achieving this is viewpoint. Firstly ... look back at the design sketches. There may be one that already has the right look and feel. If so, re-sketch it neater freehand, refine it, and use smoother linework.

Hint: By sketching the presentation drawing freehand, and being bold in the use of viewpoint and angle of lines, a more 'styley' look can be achieved, rather than a 'clunky' instrumental drawing.

Make several photocopies on cartridge paper (not thin photocopy paper) to allow for rendering with marker pens. Photocopying will also enable the drawing to be enlarged or reduced to suit. Preserve one copy as the master, on which final renderings and cut and pasted areas can be placed. Use the others for practice. Once the master is fully rendered and outlined, it can be cut out ready to be mounted onto a background.

If students have access to computers, and have the skills to use them, encourage them to use it on this drawing. James's ability on a computer speaks for itself. If hand rendering is used, the techniques must be taught — they are not a given. Precede this drawing with exercises that give practice in applying a range of media to different surface textures and form, which will ultimately inform decisions about students own designs.

The addition of a shadow and background are important elements that 'ground' the drawing.

Hint: A signature, in the right place, adds a personal and professional touch. Get students to practise writing their name in 'styley' form.

SKILLS NEEDED TO PRECEDE THIS PAGE:

Perspective drawing techniques, viewpoint and backgrounds.
Crating (proportions) and line hierarchy.
Rendering techniques — light source, tonal changes, shape and surface texture.
Colour palette considered.
Media technique: marker pens, black fineline pens, pastel pencils.
Links to: pages 24–26, 82–91

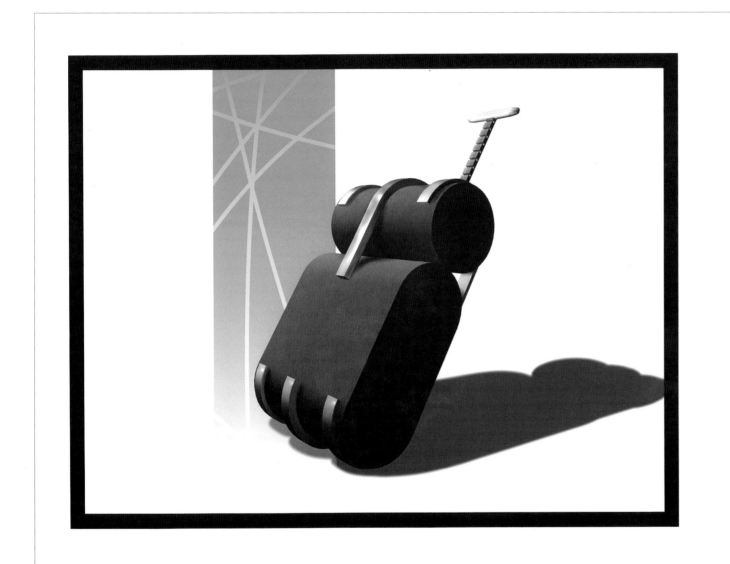

I used Maya to create a 3D model of my design, then used Photoshop to adjust lighting and shadow effects.

I would encourage other students to use a computer where possible, as it is a universal design tool.

I think my drawing is effective because I tried to keep it simple.

I added a background to give some grounding to the page, but had to be careful not to let it distract from the suitcase.

The shadow also adds a nice effect.

I made the position of the light source clear. I knew this was going to be important for when my teacher marked the drawing.

Presentation Drawing

Taylor Plank

For this drawing, I sketched my chosen design freehand. It took a while to get the proportions and the right look.

Then I got my teacher to make some photocopies for rendering. They were made a bit bigger than the sketch to get the size better to fit the page.

I rendered the different parts on different photocopies using a brown marker pen. The darker parts on the rounded edges were done with a brown pastel pencil smudged into the marker.

After rendering, I cut all the bits out and pasted them together. I found the background pattern on the internet and printed it out and pasted it onto black paper. Then I pasted my design on top and made the shadow underneath.

The parts I would change if I was to do it again is the stripe on the side. It does not look like fabric, so I would use a different technique for this. I would also use more white pen on the edges to make it stand out more.

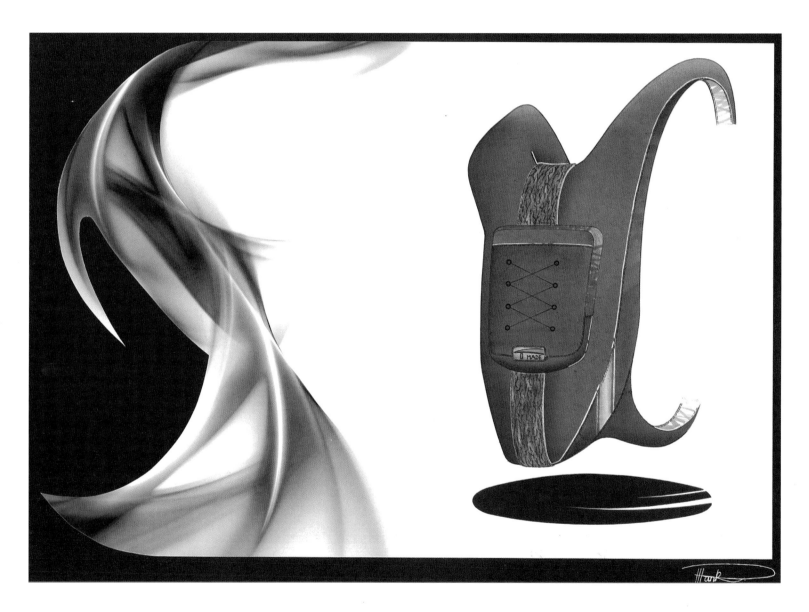

ISBN: 9780170355575

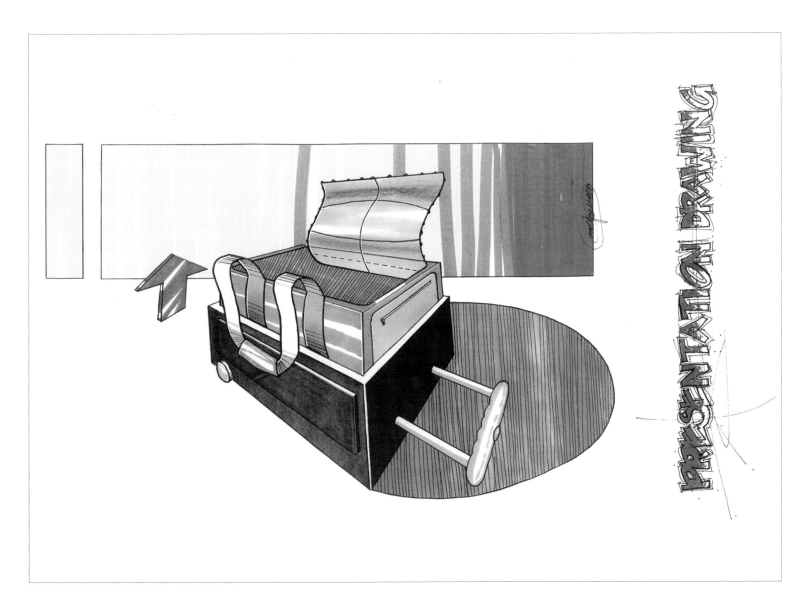

This drawing did not take as long as I thought it would because the rendering I used was mostly the same as I did on my design pages.

I did a sketch first, then got some photocopies made of it. I practised on these, and when I liked them I did them on the final copy.

The only parts that I cut out and pasted were the handle on the end and the strap in the side.

I used marker pens for all parts, and white pastel pencil smudged into the black side on the front to make it lighter than the end. I then outlined the drawing with a 0.5 black pen and cut it out.

For the background, I drew a rectangle, then used a yellow and orange marker pen to make vertical stripes through it. I then outlined it with a 0.5 black fineline pen, and cut it out and pasted it onto an A3 page.

I then pasted my rendered drawing on top and made a shadow and a title. I practised the title first so that it would look good.

ORTHOGRAPHIC PROJECTION

Preparing drawing sheets for instrumental drawings

Instrumental drawings should be set out on good quality A3 paper, with a title block. To ensure a pleasing layout on the sheet, starting points are given for most instrumental drawing exercises in this book.

Title blocks may vary but are best kept simple and placed along the bottom of the page. They should show the date, scale, the name of the drawing, the name of the person who did the drawing and the projection symbol for an orthographic drawing. The title block should demonstrate examples of the best quality lines and printing.

LINE WEIGHT
- Outlines (dark and thin) for the boundaries of the title block.
- Construction lines (light and thin) for the printing guidelines.

PENCILS
- Must be sharp.
- Use a 2H for drawing the guide lines.
- Use an HB for printing.

PRINTING
- Must be placed between 5 mm construction weight guide lines.
- Must be neat, accurate, and never rushed.
- Must use upper-case letters — avoid placing caps on the letters I and J.

The simple title block shown at right is suitable for 2D and 3D instrumental drawings. For 3D drawings (isometric and oblique, etc.) the projection symbol is not required. In this case, the section containing the student's name can be made larger.

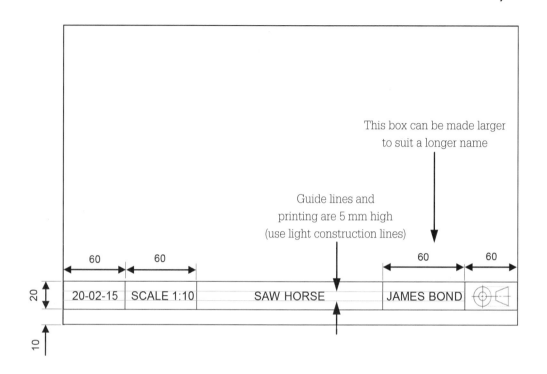

This box can be made larger to suit a longer name

Guide lines and printing are 5 mm high (use light construction lines)

20-02-15 | SCALE 1:10 | SAW HORSE | JAMES BOND

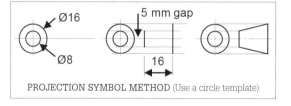

PROJECTION SYMBOL METHOD (Use a circle template)

ISBN: 9780170355575

Dimensioning in orthographic projection

The drawing below shows the correct layout for placing dimensions (sizes) around a 2D drawing (an orthographic projection). The dimensions are drawn in the same format for a plan and the elevations. Printing must be neat and enhance your drawing.

Note that where a corner is missing because of the shape of the view (at X), two short lines are drawn (the corner axes) from which to draw the extension lines.

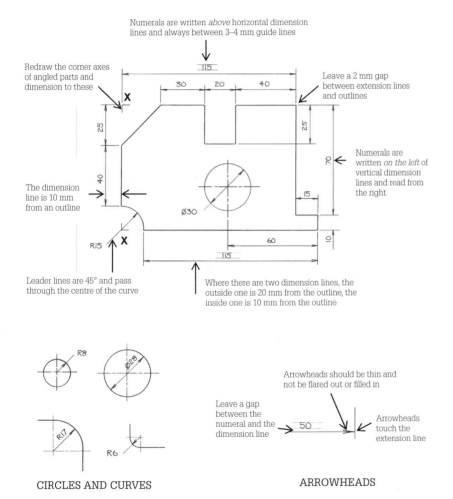

Numerals are written *above* horizontal dimension lines and always between 3–4 mm guide lines

Redraw the corner axes of angled parts and dimension to these

Leave a 2 mm gap between extension lines and outlines

Numerals are written *on the left* of vertical dimension lines and read from the right

The dimension line is 10 mm from an outline

Leader lines are 45° and pass through the centre of the curve

Where there are two dimension lines, the outside one is 20 mm from the outline, the inside one is 10 mm from the outline

CIRCLES AND CURVES

Arrowheads should be thin and not be flared out or filled in

Leave a gap between the numeral and the dimension line

Arrowheads touch the extension line

ARROWHEADS

Sectioning

Sectioning is when you cut an object open to show more clearly what the inside detail looks like when drawn in orthographic projection. Where the object has been cut is shown with a cutting plane on the drawing.

The drawings of a metal block being cut in half, and the resulting orthographic projection, are shown below. Note the following:

- How to draw the cutting plane — a centre line with two thicker lines at each end that have arrows attached, pointing in the direction of view.
- The sectioned view is always adjacent (next to) to the view that shows the cutting plane.
- Hidden detail is not shown on sectioned surfaces.
- 45° cross-hatching lines are used to identify the sectioned surfaces.
- The name of the sectioned view is written beneath it. The letters are those given to the cutting plane.

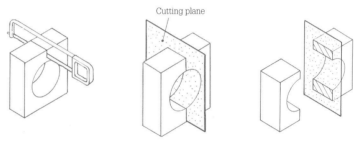

Cutting plane

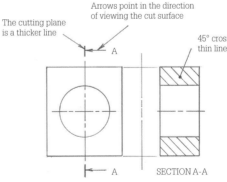

The cutting plane is a thicker line

Arrows point in the direction of viewing the cut surface

45° cross-hatching is drawn with light, thin lines. (*See next page.*)

Where a cutting plane passes through the entire object, the type of section is called a FULL SECTION

The name of the section, indicated by the cutting plane, is printed beneath the sectioned view.

A letter is given to the cutting plane

SECTION A-A

EXAMPLES OF SECTIONED OBJECTS

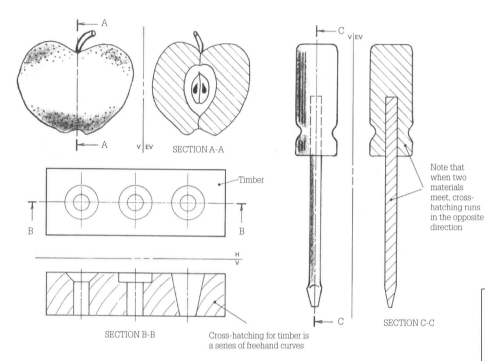

SECTION A-A

Timber

SECTION B-B

Cross-hatching for timber is
a series of freehand curves

Note that
when two
materials
meet, cross-
hatching runs
in the opposite
direction

SECTION C-C

Other types of useful sectioning
techniques are a removed and a
revolved section as seen on this
drawing of a wheel.

Removed section:
The profile (end view) of the part
is shown on a separate part of the
drawing and cross-hatched. A cutting
plane needs to be drawn to indicate
where the part has been removed
from. The view also needs to be
labelled.

Revolved section:
The profile (end view) is drawn
revolved inside the part and cross-
hatched.

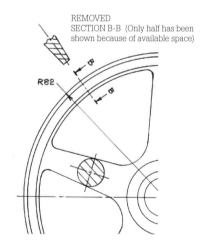

REMOVED
SECTION B-B (Only half has been
shown because of available space)

R82

HOW TO DRAW THE CROSS-HATCHING LINES FOR METAL

1 Set your dividers or compass to 3–4 mm.
2 Carefully scratch a line along the edge of your 45°
  set square.
3 Place your set square on your T-square, then
  use the scratched line as a guide to make the
  cross-hatching lines equally spaced across the
  sectioned surface.
4 Begin from the bottom of the sectioned surface
  and move upwards, placing the scratched line
  over the top of the previous drawn line.

Note: Cross-hatching lines are medium-weight
lines. They are drawn *lighter than outlines*, and are
the same weight as centre lines and hidden detail
lines. Use a sharp 2H pencil.

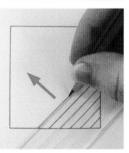

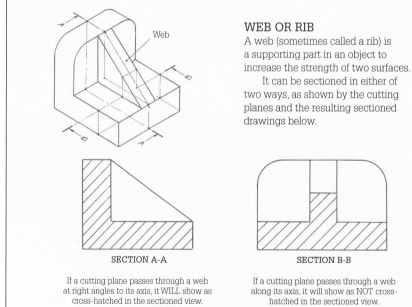

Web

WEB OR RIB
A web (sometimes called a rib) is
a supporting part in an object to
increase the strength of two surfaces.
    It can be sectioned in either of
two ways, as shown by the cutting
planes and the resulting sectioned
drawings below.

SECTION A-A

If a cutting plane passes through a web
at right angles to its axis, it WILL show as
cross-hatched in the sectioned view.

SECTION B-B

If a cutting plane passes through a web
along its axis, it will show as NOT cross-
hatched in the sectioned view.

ISBN: 9780170355575

BREAK SYMBOLS

When a part or object is too big to fit the page or drawing, it can be imagined to be cut off, called a break. Break symbols for different materials and their sectioned views are shown below.

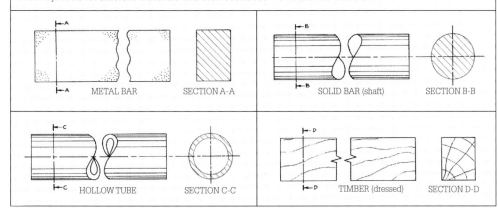

METAL BAR   SECTION A-A   SOLID BAR (shaft)   SECTION B-B

HOLLOW TUBE   SECTION C-C   TIMBER (dressed)   SECTION D-D

Note that where two adjacent materials are seen, cross-hatching runs in opposite directions.

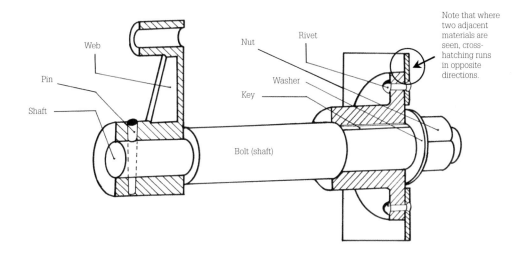

Web
Pin
Shaft
Nut
Rivet
Washer
Key
Bolt (shaft)

To make a sectioned drawing easier to read, some parts are shown not cross-hatched. This drawing shows some of the most common of these parts.

Lines

Use a sharp pencil. A 2H is ideal, or a clutch pencil with a 2H lead.

Outlines are dark and thin, all other lines are lighter and thinner.

Construction lines are the thinnest and lightest of all. These should be left on the drawing to show the methods of construction.

Remember, if the line is not an outline, it is not a dark line.

CONSTRUCTION Light and thin (the lightest of all lines) and the first lines of any drawing.	
OUTLINES Dark and thin — the last lines to be placed on a drawings after all checks and corrections have been made.	
ALL OTHER LINES Medium weight and thin: centre, hidden, reference, leader, cross-hatching, dimension lines, etc.	
BREAK LINES Medium weight and thin.	
HIDDEN DETAIL Medium weight and thin — a series of short dashes.	
CENTRE LINES Medium weight and thin — a long line broken by a short dash.	
REFERENCE LINES Medium weight and thin — a long line broken by two short dashes.	
CUTTING PLANE A centre line with a short, thicker end. There is a gap between the thick end and the centre line. Arrowheads touch the thick line and point in the direction of view.	
COMPASS CURVES Outline darkness using an HB lead in the compass. A circle template should be used for smaller curves and circles to keep the drawing neat.	

ORTHOGRAPHIC PROJECTION

EXERCISE 1: RADIO

Draw this on an A3 page, as an introduction exercise to develop the following orthographic projection skills:

- How to set out the drawing on the page and the correct line types to use.
- How to label the views, draw reference lines and the projection symbol.
- The scale of the drawing and how to write it: 1:1, 1:2, 1:5, 2:1, 1:10, etc.
- How to show dimensions.
- How to draw cutting planes and cross-hatching on sectioned surfaces.

Do the following:

1 Draw a title block on a new A3 sheet, then set up the page as shown below.
2 Using instruments, draw the radio:
- as a third angle orthographic projection (*to show four views*)
- as a paraline drawing in isometric (*looking down and at the front*)
- as a paraline drawing in oblique (*looking at the front*).
3 Use the sizes given and a scale of 1:2. Judge the sizes of any other detail.
4 Where appropriate on the drawings, show the following:
- construction lines, projection lines, outlines, centre line and any hidden detail.
- isometric circle construction, the reference line and the projection symbol.
5 Render the isometric drawing to show a reflective and a non-reflective surface.

RADIO DIMENSIONS: length 180 mm, thickness 40 mm, height 110 mm, speaker Ø80 (add any other details)

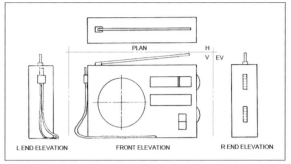

THIRD ANGLE ORTHOGRAPHIC PROJECTION

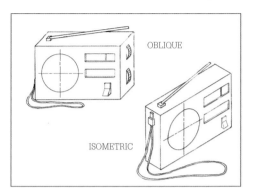

OBLIQUE

ISOMETRIC

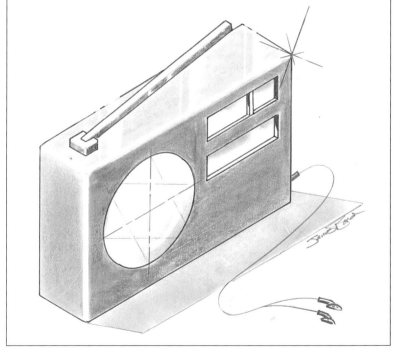

RENDERED ISOMETRIC

3rd ANGLE ORTHOGRAPHIC PROJECTION

OBLIQUE   ISOMETRIC

PAGE LAYOUT

EXERCISE 2: TOY TRUCK

The isometric drawing of a toy truck is shown.

To a scale of 1:1, redraw the truck in third angle orthographic projection. Show:

- a plan and front elevation when viewed in the direction of the arrow
- the right end elevation only
- five main dimensions
- the reference line and notation
- hidden detail and labelled views
- all projection lines.

Notes
- *The cab front slopes at 60°.*
- *Use a circle template for the wheels.*
- *Judge any sizes not given.*

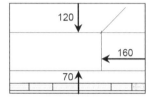

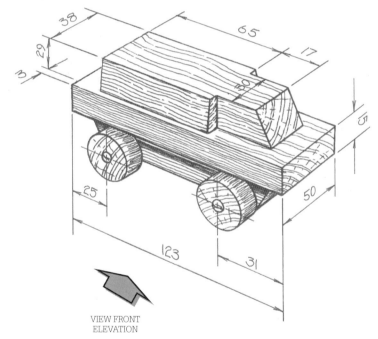

VIEW FRONT
ELEVATION

EXERCISE 3: BUILDER'S SAW HORSE

The isometric drawing of a saw horse is shown.

1 To a scale of 1:10, redraw the saw horse in third angle orthographic projection to show:
   - a plan, a front elevation and the right end elevation
   - six main dimensions and the views labelled
   - the reference line and notation
   - all hidden detail.
2 Leave projection lines and all constructions clearly visible.
3 Use a compass to construct the 75° angles of the legs.
4 Use set squares to make the inside thickness of each leg parallel to the 75° angle.
5 Produce a neat, freehand, rendered exploded isometric sketch of the wood joint found at A. Use an approximate 1:2 scale.

Hint: Draw the end elevation first (see corner X).

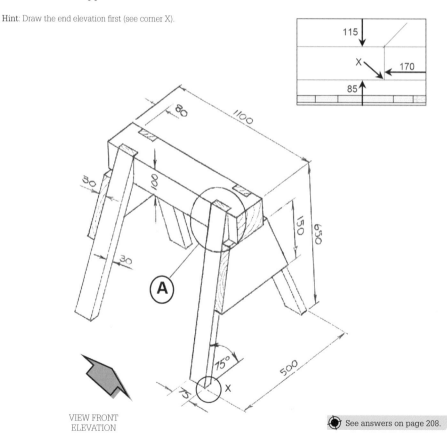

VIEW FRONT
ELEVATION

EXERCISE 4: KITCHEN SCALES

The isometric drawing and end elevation of some kitchen scales are shown.

1. To a scale of 1:2, redraw the scales in third angle orthographic projection. Show:
   - a plan, a front elevation and the left end elevation
   - four main dimensions and the views labelled
   - the reference line and notation
   - hidden detail in the end elevation only.
2. Leave projection lines clearly visible.

Hint: Draw the end elevation first (see corner X).

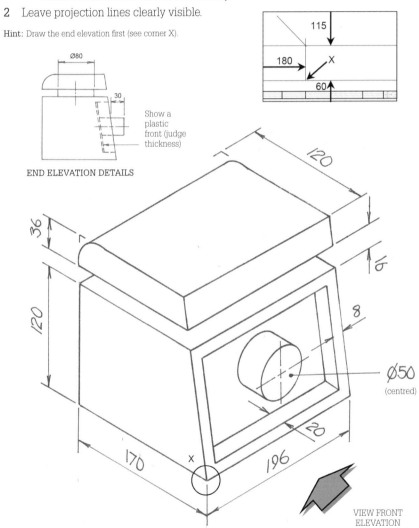

END ELEVATION DETAILS

Show a plastic front (judge thickness)

Ø80

30

36

120

120

8

20

170

196

Ø50
(centred)

16

VIEW FRONT ELEVATION

EXERCISE 5: METAL OBJECT

The isometric drawing of a metal object is shown.

1. To a scale of 1:1, redraw the object in third angle orthographic projection when viewed in the direction of the arrow. Show:
   - a plan, and the right end elevation
   - a sectioned front elevation when a vertical cutting plane passes through the central axis of the object
   - six dimensions
   - centre lines, a cutting plane, cross-hatching and hidden detail
   - the reference line and projection symbol
   - the views labelled.
2. Leave projection lines clearly visible.

Hint: Draw the centre lines first, then construct the drawing about them.

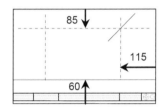

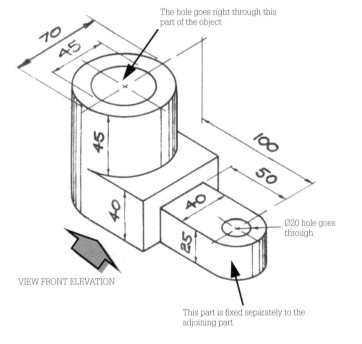

The hole goes right through this part of the object

70

45

45

100

50

40

40

25

Ø20 hole goes through

VIEW FRONT ELEVATION

This part is fixed separately to the adjoining part

85

115

60

See answers on page 209.

PARALINE DRAWING
(ISOMETRIC)

The word paraline means lines parallel to each other. A paraline drawing is a three-dimensional (3D) drawing system based on parallel lines and lines at equal angles to each other. Three surfaces of an object are shown in one drawing.

Isometric drawing

The word isometric means equal measurement: *iso* means **equal**, *metre* means **measure**. Isometric drawings are constructed around a vertical line and two sloping parallel lines, together called the **axes**. No matter how an object is viewed, the axes always retain their relationship to each other, **120°**, as shown in the cube below.

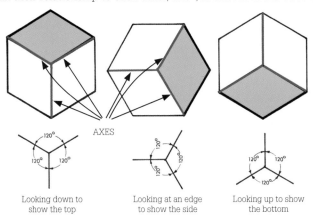

AXES

Looking down to show the top

Looking at an edge to show the side

Looking up to show the bottom

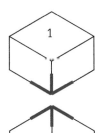

Isometric drawings are usually drawn in one of two ways:

1 Looking down to show the top. The sloping axes **slope upwards**.

2 Looking up to show the bottom. The sloping axes **slope downwards**.

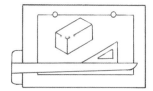

Isometric drawings are drawn with a **30°/60°** set square resting on the T-square.

Circles

ORDINATE METHOD

The word ordinate means **line**. Using ordinates is the best way to draw curves that cannot be drawn with a compass (*called compound curves — see page 67*) but it is also an accurate method of drawing a circle in isometric as shown below.

STEP 1
Draw the centre lines of the circle, then draw the circle the same size as it will be in the isometric drawing. *Use medium-weight lines.*

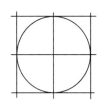

STEP 2
Draw a box (a square) around the circle. The box sides are *parallel* to the centre lines. *Use medium-weight lines.*

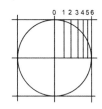

STEP 3
At random, but getting closer together towards the box, draw the ordinates. Number them. *Use medium-weight lines.*

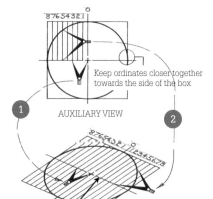

Keep ordinates closer together towards the side of the box

AUXILIARY VIEW

STEP 4
Redraw the centre lines in isometric (see over page). *Use construction lines.*

STEP 5
Redraw the box in isometric. Make the box sides *parallel* to the centre lines. *Use construction lines.*

STEP 6
Number the ordinates. Then use dividers or a sharp compass to transfer the ordinates (1) to the isometric view. *Use construction lines.*

STEP 7
Use dividers or a sharp compass to transfer (*called plotting*) the distance along each ordinate where the circle touches (2), from the auxiliary view to the isometric drawing.

Notes

- The initial box and circle with the ordinates is an extra view that needs to be drawn. It is called an *auxiliary view*.
- The number of ordinates you use depends on the size of the curve or circle, or how complicated the shape is. They do not have to be evenly spaced.
- Because the four quarters of the square and circle are the same, ordinates only need to be placed in one quarter of the circle. When transferred to isometric though, they need to be across the entire box.

Circles (cont.)
COMPASS METHOD

Although an auxiliary view of the circle or curve is not required, it is shown below to show the **right-angle rule**. The circle is drawn inside an isometric box. Compass points are located at the intersection of lines that are drawn at 90° to the side of the box where the circle will touch.

Right-angle rule: At every point where the circle will touch the box, draw a line at 90° to the side of the box.

The intersection of these lines is the compass point for drawing the curve. The radius of the curve is the distance from the point to the side of the box.

AUXILIARY VIEW

AUXILIARY VIEW

Note the right angles (90°)

ISOMETRIC
DRAWING OF A
CIRCLE

Compass points

ISOMETRIC
DRAWING OF A
CORNER CURVE

STEP 1
Draw the centre lines (opposite). Set your compass to the radius of the circle and step this distance from the middle along each line.

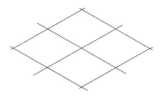

STEP 2
Draw lines *parallel* to the centre lines through the radius points of each line. This will make the box inside which the circle will be drawn.

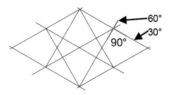

STEP 3
At each point where the centre lines touch the sides of the box, draw lines at 90° to the sides of the box. Because the box sides above are 30°, the lines will be 60° (60° + 30° = 90°).

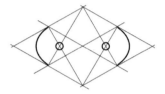

STEP 4
Where the 60° lines meet is the compass point (O). Put your compass on O, open it to the length of the line until it touches the side of the box, and draw the curves that will form the ends of the ellipse.

STEP 5
Where the 60° lines meet at the corners of the box (O) are the compass points for the top and bottom of the ellipse. Open your compass to the length of the line until it touches the side of the box, and draw the top and bottom curves. Outline the ellipse shape and place a centre line through it.

ISBN: 9780170355575

Curves can be placed on any of the sloping surfaces of an isometric or oblique drawing. The centre lines of the circle are always the first lines drawn (*drawn as construction lines to begin with*) so it is important to find out their angles to make your drawing correct. Follow the simple steps below.

1 Turn the object you want to draw into a cube. Do a quick sketch of the cube.
2 Draw the centre lines onto each surface of the cube. The centre lines are *always parallel* to the side of the cube.
3 Determine which of the surfaces your circle will be on.
4 Redraw the centre lines, with instruments, onto your isometric or oblique drawing.
5 Use the compass method or ordinates to then draw your circle.

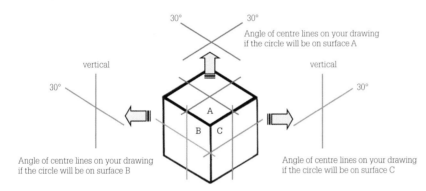

FOR ISOMETRIC DRAWINGS

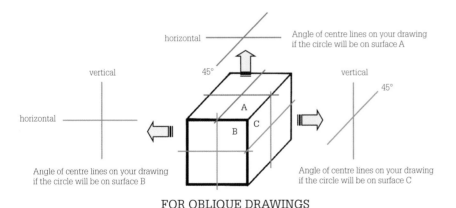

FOR OBLIQUE DRAWINGS

A cylinder

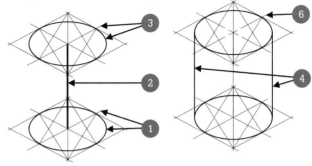

1 Draw the bottom circle. The compass method is shown.
2 Draw the vertical axis and measure the cylinder height.
3 Draw the top circle the same as the bottom one.
4 Draw the vertical sides of the cylinder.
5 Check and outline. Leave all constructions.
6 Place a centre line through the top circle.

Rounded corners

1 On the isometric axes, mark the radii of the curve.
2 Draw a right angle at these points, to intersect each other at the compass point. Set a compass to the length of the line and draw the curve.
3 To draw the bottom curve, project a vertical line from the compass point.

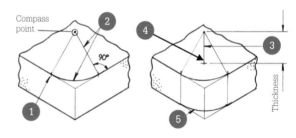

4 Mark the thickness of the object down this line to locate the new compass point.
5 Set your compass to the top radius and draw a curve from the new compass point.

Compound curves

Compound curves are free-flowing shapes that cannot be drawn with a compass. The ordinate method is used.
1 Draw the elevation of the curve and place inside a box to become an auxiliary view.
2 Divide into ordinates and transfer with dividers to the isometric drawing. A number system is a good idea.
3 Outline with a French curve or flexi curve.

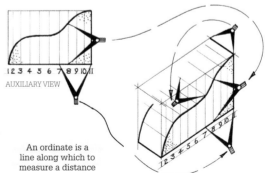

An ordinate is a line along which to measure a distance

Using an ellipse template

STEP 1
Draw the box that will contain the circle.

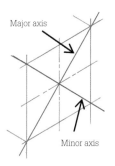

Major axis

Minor axis

STEP 2
Identify the major and minor axes for the ellipse. The major axis will be at 60°, the minor axis 30°. (They will also be at 90° to each other.)

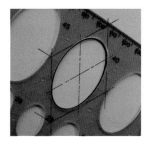

STEP 3
Line up your ellipse template with the major and minor axes and draw the ellipse.

Exploded isometric drawings

Exploded isometric drawings are useful to show more clearly how the parts of an object are fitted together. All parts are shown pulled out along their central axes (exploded) in the same direction in which they are assembled. If they were pushed together they would fit as intended and form an isometric drawing.

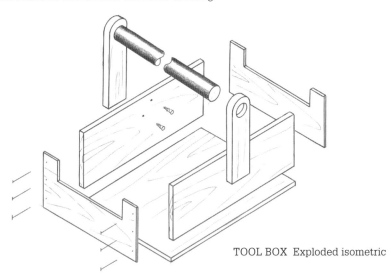

TOOL BOX Exploded isometric

Isometric dimensioning

The drawing below shows the correct layout for placing dimensions (sizes) about a 3D isometric drawing. There are more dimensions than necessary to show how they can be placed on all surfaces.

RULES
- Dimension lines are **10 mm** from the outlines.
- Numbers are always placed between **3–4 mm** guide lines with a sharp **HB pencil**.
- Guidelines are drawn lightly with a **2H pencil**.
- Extension lines do not touch the drawing. Leave a **2 mm** space between the object.
- Arrowheads are thin and should not be flared out or filled in. The width of an arrowhead should be approximately **one third of its length**.
- Numbers face the viewer and are read in a clockwise direction. Numbers are on the **outside** of the dimension line unless they are at the bottom of an object.
- Circles are stated as a radius or diameter, depending upon available space and the size of the circle.
- Your printing of the numbers must be very neat and enhance your drawing.

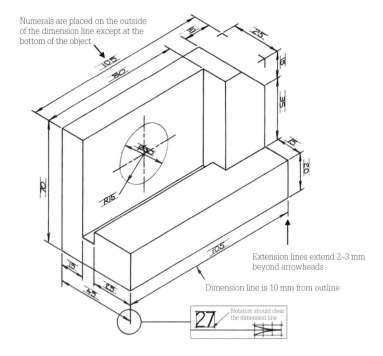

Numerals are placed on the outside of the dimension line except at the bottom of the object

Extension lines extend 2–3 mm beyond arrowheads

Dimension line is 10 mm from outline

Notation should clear the dimension line

ISBN: 9780170355575

ISOMETRIC DRAWING

Do these exercises to develop the following isometric drawing skills:

- Line types — construction, outlines, centre lines.
- How to draw circles and curves.
- How to draw objects when looking down on them and/or up at them from beneath.
- The scale of the drawing and how to write it: 1:1, 1:2, 1:5, 2:1, 1:10, etc.
- How to show dimensions (the sizes of the object).
- How to draw exploded isometric views.

EXERCISE 1: SHAPED BLOCK

The orthographic projection of a shaped block is shown.

1 Use a scale of 1:4 to redraw the block in isometric when viewed in the direction of the arrow.
2 Show six dimensions and all constructions lightly and clearly.

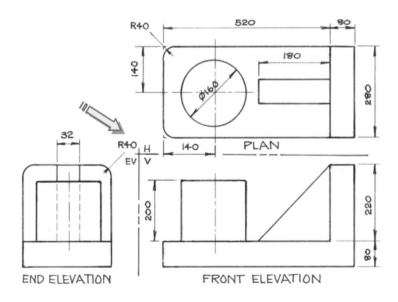

EXERCISE 2: THEATRE SPOTLIGHT

The orthographic views of a theatre spotlight are shown.

1 Use a scale of 1:2 to redraw the spotlight in isometric when viewed in the direction of the arrow.
2 Show four dimensions and all constructions lightly and clearly.
3 Produce two concept sketches of a bracket design to hold the spotlight.
4 Show your chosen design on the isometric drawing.
5 Render the spotlight to show form (*shape and surface qualities*).

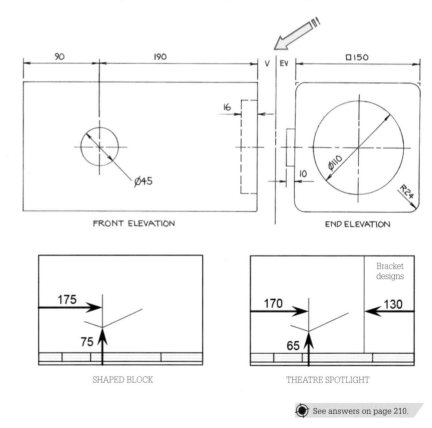

See answers on page 210.

EXERCISE 3: EXPLODED ISOMETRIC WOOD JOINT

The isometric drawing of a wood joint is shown.

1 Use a scale of 1:1 to redraw the block in exploded isometric when pulled apart in the direction of the arrow.

2 Colour render the drawing to show timber.

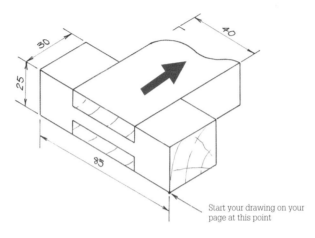

Start your drawing on your page at this point

Method

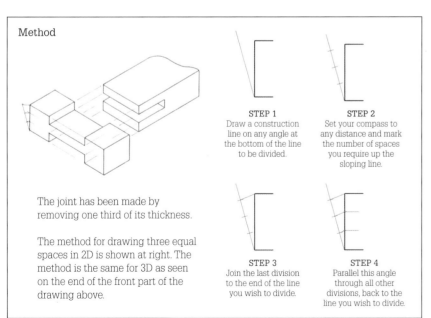

The joint has been made by removing one third of its thickness.

The method for drawing three equal spaces in 2D is shown at right. The method is the same for 3D as seen on the end of the front part of the drawing above.

STEP 1
Draw a construction line on any angle at the bottom of the line to be divided.

STEP 2
Set your compass to any distance and mark the number of spaces you require up the sloping line.

STEP 3
Join the last division to the end of the line you wish to divide.

STEP 4
Parallel this angle through all other divisions, back to the line you wish to divide.

EXERCISE 4: EXPLODED ISOMETRIC TOY TRAIN

The isometric drawing of a toy train is given below. Take sizes directly from it to produce an exploded isometric drawing. Follow the steps shown in the photos to draw the wheels using an ellipse template. Use the template to also draw the funnel and the engine.

STEP 1
Draw the major and minor axes (60° and 30°) and draw a half ellipse for the back of the wheel.

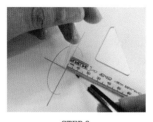

STEP 2
Step out the thickness of the wheel along the minor axis.

STEP 3
Draw another major axis, then draw the front of the wheel, the same size as the back.

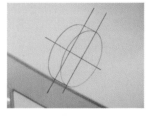

STEP 4
Use your 30° set square to join the two ellipses to make the top and bottom of the wheel. Erase lines not needed.

TIP

To make an exploded drawing easier to read, leave a gap between the exploded parts.

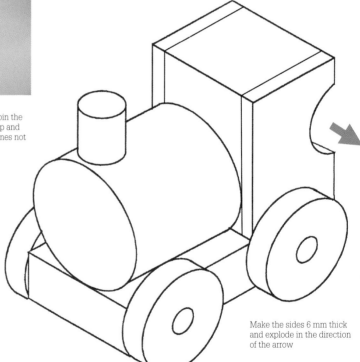

Make the sides 6 mm thick and explode in the direction of the arrow

See answers on page 210.

Parts list

A parts list makes understanding a drawing easier. It shows clearly the name of the part, how many of each part, and what material the part is made from.

Parts lists are common on exploded isometric drawings, but can also be applied to any instrumental drawing to clarify understanding.

An exploded isometric drawing of a coffee maker with a parts list is shown.

Note the following:

The parts list

- The parts list is attached to the title block at the bottom of the page and is read from the bottom upwards.
- If attached to the top of the page, the columns read downwards.
- Printing is 4 mm high (use guide lines).
- Each row is 6 mm wide.
- There are four columns.
- It can be made on a computer and pasted onto the drawing. A clean font, such as Arial, is best, with a size approximately 9.

On the drawing

- About the drawing, neatly print the number of each part of the drawing, to match the number you have given it in the item column of the parts list.
- Place a circle around the number. *(Use a circle template.)*
- Place a leader line from the *centre* of the circle. *(A medium-weight line.)*
- If the leader line touches the part, it will end with an arrowhead.
- If the leader line extends inside the boundary of the part, it will end with a bullet or dot.

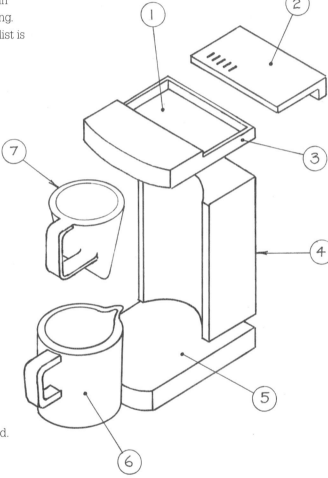

ITEM	PART NAME	QTY	MATERIAL
1	WATER HOLDER	1	PLASTIC
2	LID	1	PLASTIC
3	TOP	1	PLASTIC
4	MAIN BODY	1	METAL
5	BASE	1	METAL
6	JUG	1	STAINLESS STEEL
7	FILTER	1	PLASTIC

Note that if there is not room at the bottom of the page, the parts list can be placed at the top as seen here. In this case, the columns read downwards.

Leader lines look best when they begin at the centre of the circle.

7	FILTER	1	PLASTIC
6	JUG	1	STAINLESS STEEL
5	BASE	1	METAL
4	MAIN BODY	1	METAL
3	TOP	1	PLASTIC
2	LID	1	PLASTIC
1	WATER HOLDER	1	PLASTIC
ITEM	PART NAME	QTY	MATERIAL

20-5-15	SCALE 1:2	COFFEE MAKER	JAMES BOND

Oblique drawing

The word oblique means **angle**. Oblique is a pictorial or 3D drawing method showing three surfaces of an object in one drawing. Oblique drawings are constructed around a vertical, a horizontal, and one sloping line called the **axes**. Oblique drawings show the **true shape** of the surface closest to the viewer. This means that the surface is **parallel** to the line of sight of the viewer.

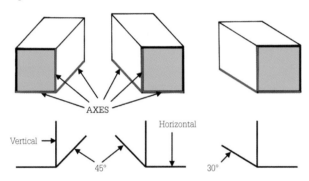

The sloping axes are drawn at 45°. A 30° angle may be used to give a less distorted look, but 45° is more commonly used to prevent confusion with isometric drawings.

There are two types of oblique drawings:

Cavalier oblique — all axes are drawn to the given sizes, i.e. **full size.**

CAVALIER OBLIQUE

CABINET OBLIQUE

Cabinet oblique — the horizontal and vertical axes are drawn full size, the sloping axis is drawn **half size.**

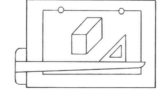

Oblique drawings are drawn with a **45°** set square resting on the T-square.

Oblique circles

TRUE SHAPE METHOD

The true shape of an object or surface is seen when you look directly at it.

    True shape circles are seen when their 3D shape is a cylinder and when you are looking directly at the circular end.

1   Draw the back circle centre line and draw the circle.
2   Draw the 45° axis (length) of the cylinder.
3   Draw the front circle centre line and draw the circle.
4   Join the circles with 45° lines to form the cylinder sides.
5   Place a centre line through the front circle.

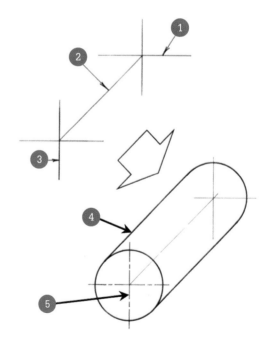

COMPASS METHOD

This method is used for circles that are seen on the sloping side of an object. It is the same method as used for isometric circles.

1 Draw the centre lines of the circle. *(The slope of the centre lines are parallel to the sides of the surface on which the circle will be placed. The circle shown is for a vertical surface.)* (See page 67.)

2 From the centre, step out the radius of the circle along the centre lines. *(Use a compass.)*

3 Draw a box whose sides are parallel to the centre lines.

4 Draw right angles to the sides of the box at every point on the box where the circle will touch.

5 At the intersection of these lines, set your compass to the length of the line and draw the circle in four parts.

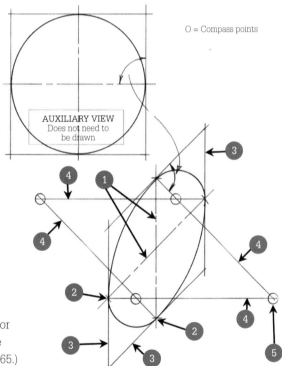

O = Compass points

AUXILIARY VIEW
Does not need to be drawn

You can also use ordinates for drawing oblique circles. See isometric circles. (See page 65.)

Oblique dimensioning

The drawings below show the correct layout for placing dimensions (sizes) around a 3D oblique drawing. There are more dimensions than necessary to show how they can be placed on all surfaces.

Two cubes are shown.

- Cube 1 shows all dimensions.
- Cube 2 shows how the dimensions at the back of a drawing may be arranged as an alternative to cube 1.

RULES

- The same rules apply as for orthographic projection.
- Dimension lines are **10 mm** from the outlines.
- Numbers are always placed between **3–4 mm** guide lines with a sharp **HB pencil**.
- Guidelines are drawn lightly with a **2H pencil**.
- Numbers are **above** horizontal dimension lines and on the **left** of vertical dimension lines, read from the right side. (See alternative positioning of numerals at back of objects below.)
- Extension lines do not touch the drawing. Leave a **2 mm** space between the object.
- Arrowheads are thin and should not be flared out. The width of an arrowhead should be approximately **one third of its length**.
- Circles are stated as a radius or diameter, depending upon available space and the size of the circle. Leader lines are at 45°.
- Your printing of the numbers must be very neat and enhance your drawing.

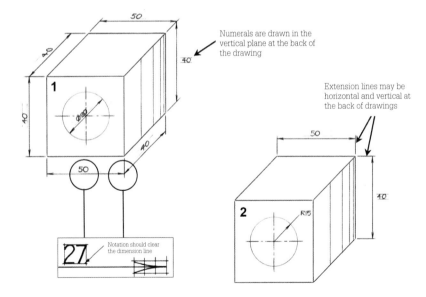

Numerals are drawn in the vertical plane at the back of the drawing

Extension lines may be horizontal and vertical at the back of drawings

Notation should clear the dimension line

OBLIQUE DRAWING

Do these exercises to develop the following oblique drawing skills:

- Line types — construction, outlines, centre lines.
- How to draw circles and curves and use an auxiliary view to plot angles.
- How to draw objects in cavalier and cabinet oblique.
- The scale of the drawing and how to write it: 1:1, 1:2, 1:5, 2:1, 1:10, etc.
- How to show dimensions (the sizes of the object).
- How to plot an angle using an auxiliary view.

EXERCISE 1: RADIO

The orthographic projection of a radio is shown.

1   Use a scale of 1:2 to redraw the radio in cavalier oblique when Face A is true shape (*facing towards you, looking in the direction of the arrows*).

2   Show four dimensions and all constructions.

EXERCISE 2: CAST IRON STAND

The orthographic projection of a cast iron stand is shown.

1   Use a scale of 1:2 to redraw the stand in cavalier oblique when Face C is true shape (*facing towards you*).

2   Show four dimensions and all constructions.

Note: An auxiliary view is required to draw the 75° angle of the sloping sides in oblique. Show this lightly but clearly on the page.

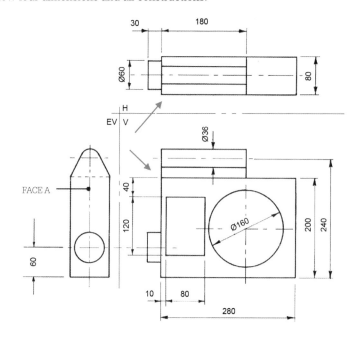

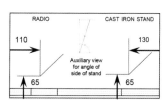

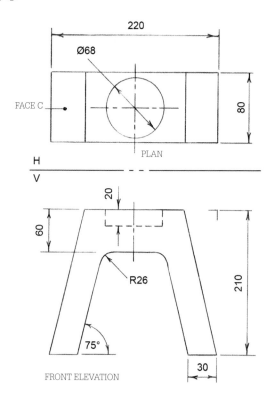

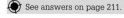
See answers on page 211.

PRODUCT DESIGN

Design sketching

To make your design sketches more exciting to look at — to 'pop' on the page — and to show your designs to advantage, use these simple techniques.

Line weight

Using lines of different thickness (line hierarchy) will help clarify a form and add a dynamic feel to the sketch.

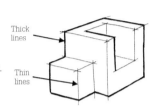

- **Thick lines** are those that join two surfaces but where only one of the surfaces is seen.
- **Thin lines** are those that join two surfaces but where both surfaces are seen.

Shadows

Shadows can be a cast shadow from a light source or a simple drop shadow underneath the object. Whichever you choose, a shadow will give your sketch life and make it look more realistic. Use a 4B pencil smudged, light grey marker pen, or a series of vertical lines, to shade the shadow.

Backgrounds

A background 'box' is a quick way of tightening up a sketch to make it stand out. The box can be horizontal or vertical. A background can also be used to tie a range of designs together. Backgrounds can be lightly rendered — pastel pencil smudged with your finger from the edge of a piece of paper works well, or a light grey marker pen.

 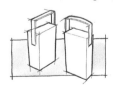

Viewpoint

Give clearer understanding of your designs by sketching your ideas when looking from different directions. For example, sketch when looking up from a low point (on the ground), when looking down onto the top, when looking at the corner or side or at the back. Make the crates on steeper angles for a more dynamic look.

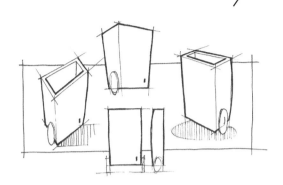

Quick rendering

Don't know where to start? Keep it simple. Use markers if you have them.

1. Outline the sketches in black pen using thick and thin lines.
2. Choose where the light is shining from, then colour the areas furthest from the light.
3. Think about shiny and dull surfaces. Top shiny surfaces will have a vertical reflection which can be done with an eraser against the edge of a piece of paper. These steps may be all that is needed to convey the form of the design.

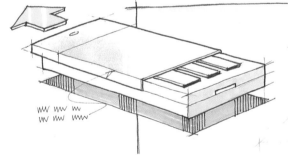

Remember to have three tones: light on the top surfaces, darker on one side and darkest on the surfaces furthest from the light source. Colours should indicate the material — blue for metal, brown for wood, etc.

ISBN: 9780170355575

Planning page

Before beginning ideation sketches, it is a good idea to plan each page on a sheet like the one below. A planning page will allow you to decide on the following:

- How many pages will be in the set and what each page will look like.
- How many sketches to show, what crating methods to use, and the proportions of each.

- What types of sketching methods, viewpoints, and how many will be on each page.
- What will be the colour palette for each page.
- What types of backgrounds and shadows work best.
- Where titles, research and notes will be placed.

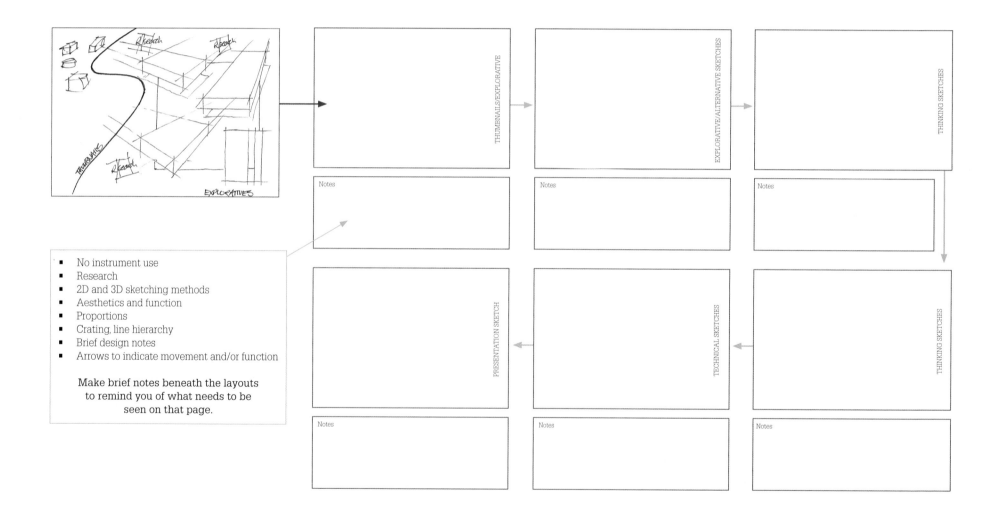

- No instrument use
- Research
- 2D and 3D sketching methods
- Aesthetics and function
- Proportions
- Crating, line hierarchy
- Brief design notes
- Arrows to indicate movement and/or function

Make brief notes beneath the layouts to remind you of what needs to be seen on that page.

ISBN: 9780170355575

Handwritten titles

To clarify information in a set of design drawings, the addition of a title on each page should be considered. Although the title may be made on a computer and pasted onto the page, a handwritten title furthers the development of a dynamic style and complements the freedon of ideation sketches. It is important that the title enhances the page, and is not merely an 'afterthought' that has been rushed. It should be neat, with a well-considered lettering style (font). The placement of the title on the page is also important.

As a guide, use block, upper-case letters, placed between 10–12 mm guide lines.

Do this exercise to develop the following handwritten title skills:
- A font style and proportions of each letter — spaces between for clarity.
- What types of lines to use — thick for the letters, thin for the border around each letter.
- What media to use — pencil, black fineline pens, marker pens.

1 On a piece of A4 paper, lay out two sets of guide lines 12 mm apart. (Use a straight edge and a soft pencil so that they can be erased later.)
2 Carefully construct each letter of the alphabet the same or similar to the ones shown below (take care when forming each letter to ensure a pleasing look).
3 When you are happy, outline the block letters using a 0.3 black fineline pen.
4 Use a 0.1 black fineline pen to place a thinner border around each letter.
5 Render with marker pen. (For a better look, only colour the top or bottom half of each letter.)

STEP 1
Draw the letters in soft pencil first, between 12 mm guide lines.
Use a 0.3 black fineline pen to outline each letter.

STEP 2
Use a 0.1 black fineline pen to place a border around each letter.

STEP 3
Choose a colour to match the colour palette of the page and use a marker pen to render the letters.

Caroline Webb Year 10

- Placement on the page is important. Titles can be horizontal, vertical or curvy.
- Make vertical titles read from the right side of the page.
- For effect, render only the top or bottom half of each letter.
- Dots and other well-considered 'scribbles' with a fineline pen provide a more expressive look.

Do this exercise to develop the following 3D freehand sketching skills:

- Proportions and viewpoints: *place sketches in crates.*
- How to construct circular parts.
- What types of lines to use: line hierarchy (*thick and thin lines, dark and light lines*).
- What pencil to sketch with: *soft pencil* (*HB or 4B*).
- What media to render with: *marker pens, pastel pencil, fineline pen, 4B pencil.*
- How to quick render: *to show the tonal qualities produced by an identified light source and its three-dimensional effects on an object. Show light and dark surfaces.*
- Shadows and backgrounds.

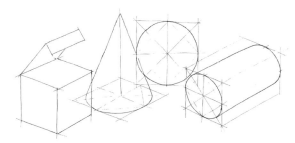

STEP 1
Draw the four solids above inside correctly proportioned crates. Note the circle construction method for the sphere, the bottom of the cone and the end of the cylinder. *Make the angles of the cube, cone and cylinder low, not steep.*

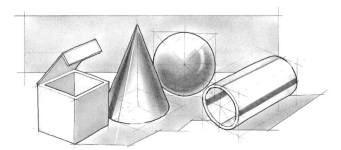

STEP 2
Turn the cube into an empty box with side thickness lines and make the cylinder hollow. Use marker pen to render the cube and cylinder, and smudged 4B pencil for the cone, sphere and shadow. Use smudged pastel pencil for the background.

Do this exercise to develop the following product design drawing skills:

- How to plan each page.
- How to consider viewpoint.
- How to sketch in 2D and 3D.
- How to show crating, line hierarchy and quick rendering.
- How to produce hand-drawn titles.
- How to show human factors, and think about alternative ideas.
- How to sketch exploded detail that shows construction of the product.
- How to draw a rendered presentation drawing.

Note: To prepare students for a product design assignment, this practice brief must be led by the teacher to show the class the layout and skills required for each page. Students may add their own design ideas. The work should not be rushed. Time: approximately three to four weeks.

EXERCISE 1: MOBILE PHONE

SITUATION
Mobile phones are now a part of daily life. A company has asked you to design a new-look phone.

TASK
Design a new mobile phone.

SPECIFICATIONS
The phone must consider aesthetics, function and human factors.

REQUIREMENTS
1   From your planning sheet, work on your own A3 paper to produce the following set of drawings:
    PAGE 1: Thumbnails and explorative sketches. Include research.
    PAGE 2: Thinking sketches. Include alternative ideas and human factors.
    PAGE 3: Technical sketches and a presentation sketch.
2   Make hand-drawn title blocks for all the sheets.
3   Present the drawings stapled together, down the left side, landscape format, with a front cover that you have made. *This may be computer generated.*

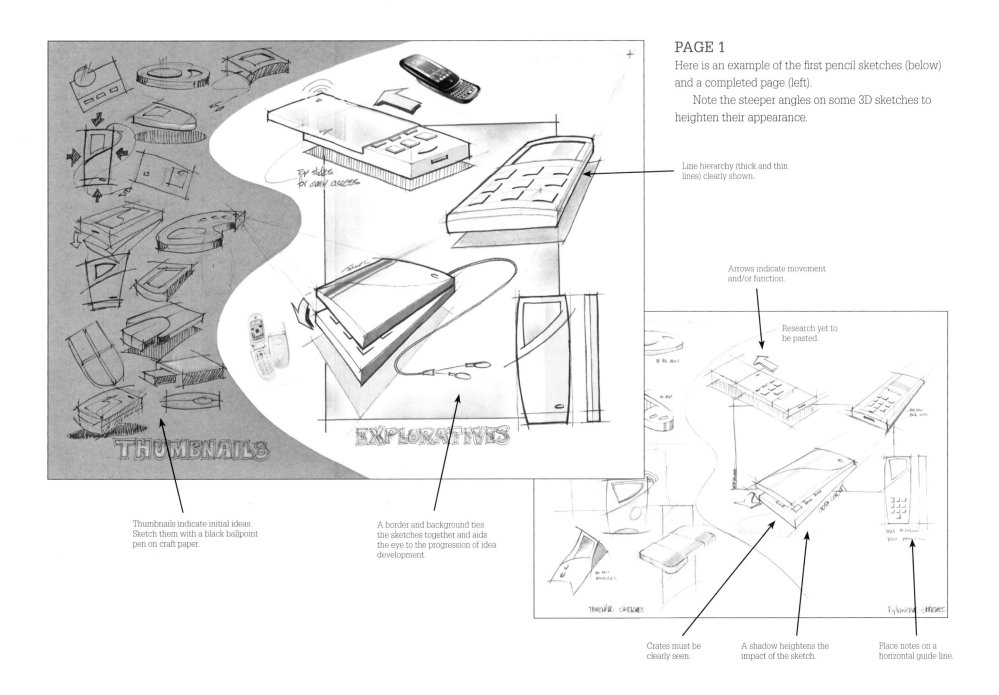

PAGE 1

Here is an example of the first pencil sketches (below) and a completed page (left).

Note the steeper angles on some 3D sketches to heighten their appearance.

Line hierarchy (thick and thin lines) clearly shown.

Arrows indicate movement and/or function.

Research yet to be pasted.

Thumbnails indicate initial ideas. Sketch them with a black ballpoint pen on craft paper.

A border and background ties the sketches together and aids the eye to the progression of idea development.

Crates must be clearly seen.

A shadow heightens the impact of the sketch.

Place notes on a horizontal guide line.

THUMBNAILS

EXPLORATIVES

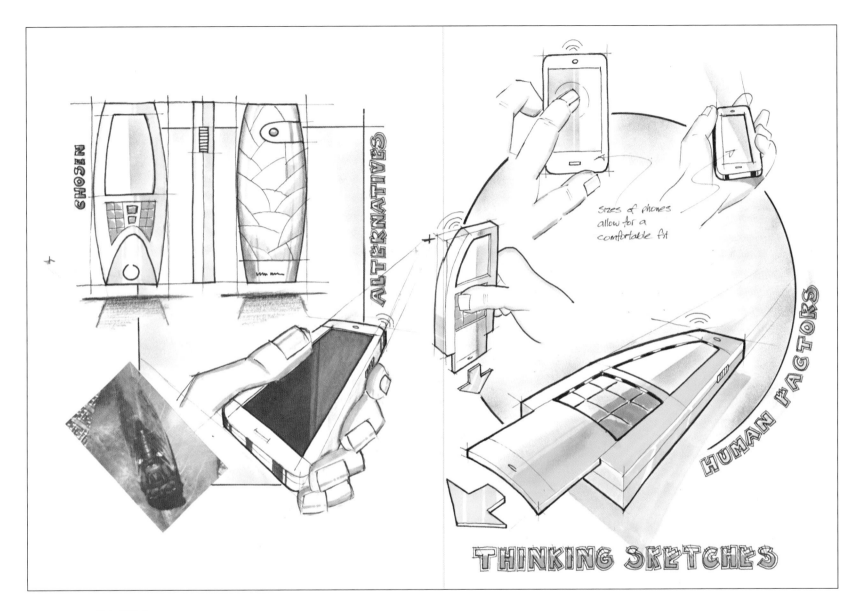

PAGE 2

Show: 2D and 3D sketches, variety of viewpoints, crating, line hierarchy, quick rendering to show shape, texture and materials, arrows to show movement and/or function, alternative ideas, human factors, notes, research, backgrounds, shadows, hand-drawn titles.

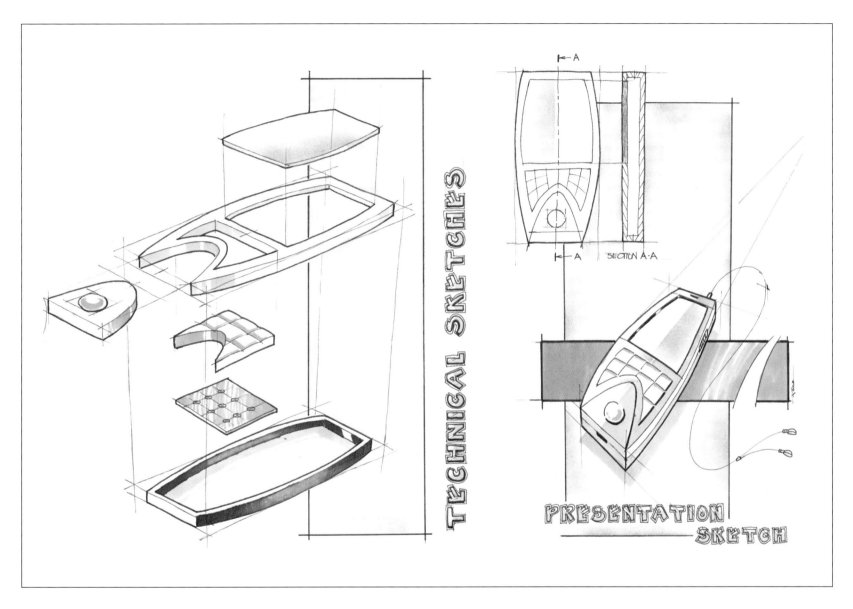

TECHNICAL SKETCHES

SECTION A-A

PRESENTATION SKETCH

PAGE 3

Show: 2D and 3D sketches, variety of viewpoints, crating, line hierarchy, quick rendering to show shape, texture and materials, exploded isometric sketches, 2D sectioned views and a cutting plane and cross-hatching, notes, research, backgrounds, shadows, hand-drawn titles.

EXERCISE 2: RENDERING

Do this exercise to develop the following rendering skills for a product design presentation drawing:

- Determining a light source.
- Use of rendering products (media), particularly marker pen.
- Tonal changes and surface qualities.
- White highlights, reflections and shadow.
- Pastel background on a 'toothed' mounting card.

Note: The completed drawing is shown below. An outline drawing is shown at right. Redraw this and make four photocopies on cartridge paper, then follow the steps shown in the photographs on the following pages to render the phone. Do not rush it.

A wide range of rendering products is required:

- Media: *marker pens, black fineline pens (0.1 and 0.3), white pastel pencil, white rollerball highlight pen, 4B pencil, gouache.*
- Equipment: *French curves, eraser, erasing shield, straight edge, spray-mount glue, craft knife.*

Time: one week.

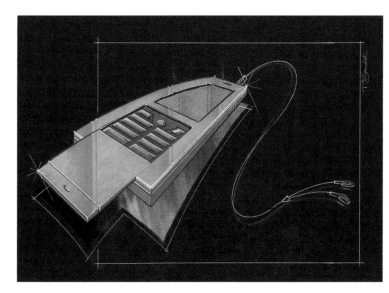

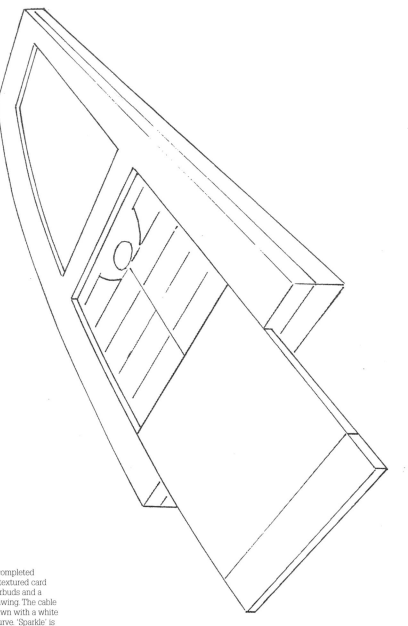

Here is an example of the completed rendering mounted onto a textured card backing. The addition of earbuds and a signature enhances the drawing. The cable and earbuds have been drawn with a white gel pen against a French curve. 'Sparkle' is applied with gouache and a white gel pen.

ISBN: 9780170355575   PHOTOCOPYING OF THIS PAGE IS RESTRICTED UNDER LAW.

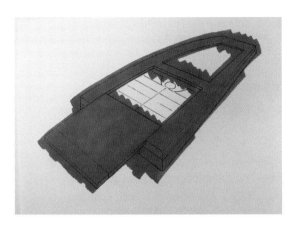

STEP 1

- Cover entire body surface with smooth strokes of a marker pen. (*A darker colour is best.*)
- Make direction of strokes parallel to the longest axis.
- Leave to dry thoroughly.

STEP 4

- Use an eraser against an erasing shield to remove random amounts of white pastel.
- A series of wide and thin stripes will create shine on the top surface.
- The stripes must be vertical. (*Light forms vertical stripes on a horizontal surface.*)

STEP 2

When the marker is thoroughly dry, go over the top surface with a liberal application of white pastel pencil.

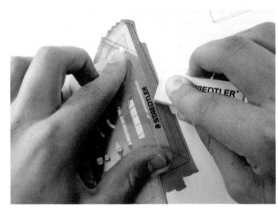

STEP 5

Use an eraser against an erasing shield to remove the pastel that has smudged onto the edges of the object.

STEP 3

- Use your finger tip to smudge the pastel into the marker.
- Circular strokes are best to achieve a smooth texture.

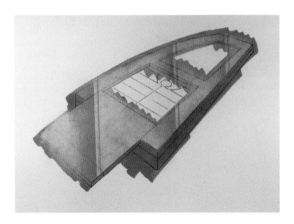

STEP 6

The rendering should now look like this.

ISBN: 9780170355575   PHOTOCOPYING OF THIS PAGE IS RESTRICTED UNDER LAW.

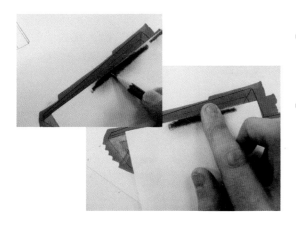

STEP 7

- Apply tonal value to the side by using 4B pencil smudged from the edge of a piece of paper.
- This can be applied to all surfaces that would be in shadow.

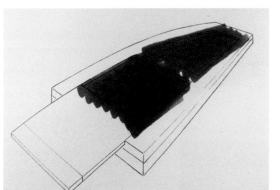

STEP 8

- On another photocopy, cover the screen and keypad with smooth strokes of a marker pen. (*Dark grey C08.*)
- Make direction of strokes parallel to the longest axis.
- Leave to dry thoroughly.

STEP 9

When the marker is thoroughly dry, apply a liberal application of white pastel pencil.

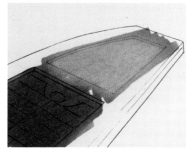

STEP 10

Use your fingertip to smudge the pastel into the marker. Create reflection with vertical eraser stripes against an erasing shield.

STEP 11

- Outline the shape of the screen with a fineline black pen.
- Fill in the edge of the screen with black pen to create depth (above).
- Use an ellipse template for the curved side.

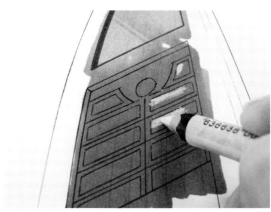

STEP 12

- Apply white pastel pencil to highlight the tops of each key on the keypad.
- Smudge the pastel for a smooth effect.

ISBN: 9780170355575

STEP 13

Outline the features of the keypad with black fineline pen. Apply white highlight pen to the top edges. *Use a straight edge.*

STEP 16

- Cut and paste the screen and keypad onto the body of the phone.
- Carefully cut out the completed phone and place onto a piece of clean paper to experiment with a background.

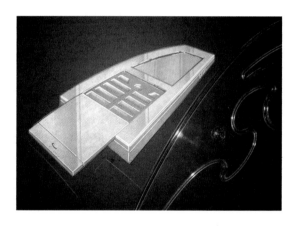

STEP 14

- Highlight features and edges of the main body with a white roller-ball highlight pen.
- Use a straight edge.

STEP 17

- Redraw the chosen background onto the mounting card.
- The mounting card should have a 'tooth' that will allow for the successful application of pastel.

STEP 15

Outline the profile line of the phone with a black 0.3 fineline marker pen.

STEP 18

- Apply a layer of Frisket Film onto the mounting card.
- With a sharp craft knife, make a mask by carefully cutting out the shape of the background.

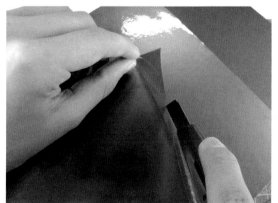

ISBN: 9780170355575

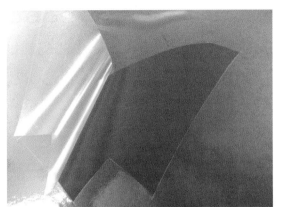

STEP 19
Peel off the Frisket Film to reveal the background shape.

STEP 22
- Remove the mask and all unwanted pastel.
- Added definition to the background can be achieved by adding a drop shadow with a white highlight pen.
- Note that the background will need to be 'fixed' with a spray fixative.

STEP 20
Using a craft knife against a chalk pastel stick, scrape a range of suitable colours onto the edge of the mask.

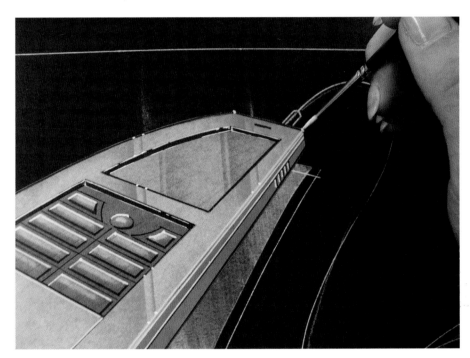

STEP 21
- Smudge the pastel into the background with a tissue.
- Use curved strokes for added interest in the final background.

STEP 23
Spray-mount the phone onto the background. Add 'sparkle' with gouache, applied with the tip of a thin brush. Final touches such as the addition of headphones, controls and reflection into the background indicate function and enhance the final presentation. Sign the final drawing with a 'styley' signature. Practise it first.

ISBN: 9780170355575

Here is an alternative mobile phone shape that introduces curved surfaces. Redraw it and make four photocopies as in the previous exercise and then follow the steps shown in the photographs.

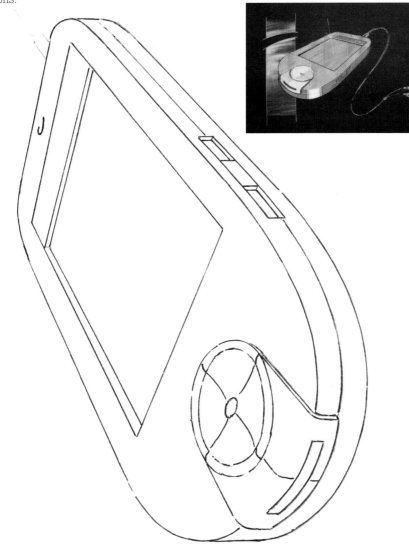

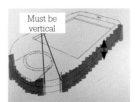

Make marker strokes vertical as shown by the arrow, leaving the rounded front corner white. The thin lines must also be vertical and parallel to each other.

When the marker is thoroughly dry, apply pastel pencil of the same colour on top of the marker at the rounded front corner.

Blend the pastel into the white area by smudging with vertical strokes of your finger across the width of the side. Leave the middle part white.

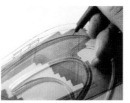

Use a 0.1 fineline black pen to outline the top part of the phone. Use a French curve for the curved parts and a set square for the straight parts.

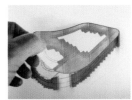

Carefully cut out the top part against, and leaving, the black outline. Spray-glue it onto the side part. Take care to get it in the right place first time.

Carefully cut out the screen against, and leaving, a black outline. Spray-glue it onto the top. Take care to get it in the right place first time.

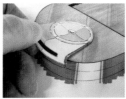

Render the controls and outline with a 0.1 fineline black pen. Then carefully cut out against, and leaving, the black outline, and spray-glue onto the top of the phone.

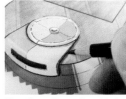

Use either a 4B pencil or a black colouring pencil to add shadow detail around the controls.

Use a white pastel pencil to create highlight on the rounded edge. Leaving a white strip when rendering makes this easier.

Outline with a 0.1 fineline black pen. Use a white highlight pen to place a white line beside the black line to highlight features and edges. Use your pen against a French curve and a set square. **Do no lines freehand.**

Cut the drawing out and place it onto a background page. Draw a simple backgound and use a French curve to draw the earbud cord. Render the background then spray-glue the phone on top. **Important:** Make sure the vertical edges are vertical when gluing the phone onto the background.

EXERCISE 3: MATERIALS RENDERING

Do these exercises to develop a range of rendering techniques for materials commonly found in product design.

Storage Box

Material: *polished wood*

Media: *colouring pencils, HB pencil, black fineline pen, grey marker pen*

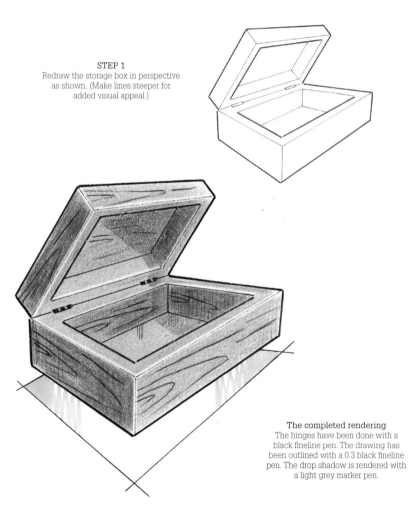

STEP 1
Redraw the storage box in perspective as shown. (Make lines steeper for added visual appeal.)

The completed rendering
The hinges have been done with a black fineline pen. The drawing has been outlined with a 0.3 black fineline pen. The drop shadow is rendered with a light grey marker pen.

STEP 2
Apply brown colouring pencil to the sides furthest from the light source. Make your hand movement parallel to the longest edges.

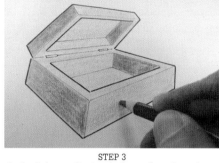

STEP 3
Apply a lighter application of colour to the surfaces closest to the light source. For a polished wood look, apply orange and yellow colouring pencil on top of the brown surfaces.

STEP 4
Wrap your finger in a tissue and smudge the colour into all surfaces.

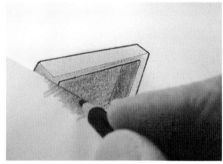

STEP 5
Make a shadow with a light application of black colouring pencil from a paper edge.

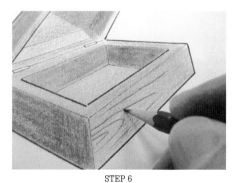

STEP 6
Draw wood grain with a sharp HB pencil. Don't make them too curvy, mostly straight, following the angle of the sides.

STEP 7
Use a sharp eraser against an erasing shield to make white edges.

ISBN: 9780170355575

Satchel

Material: *textiles*

Media: *pastel pencils, colouring pencils, marker pens, 0.3 black fineline pen, white gel pen*

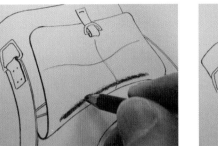

STEP 2
Apply brown pastel pencil to the raised areas.

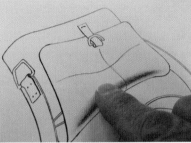

STEP 3
Smudge the pastel with your finger.

STEP 4
Use your eraser to make the rounded edges.

STEP 1
Redraw the outline of the satchel.
(Use a french curve.)

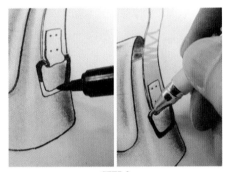

STEP 5
Render the strap with marker pen.

STEP 6
Render the buckles with a black marker pen.
Add highlights with a white gel pen.

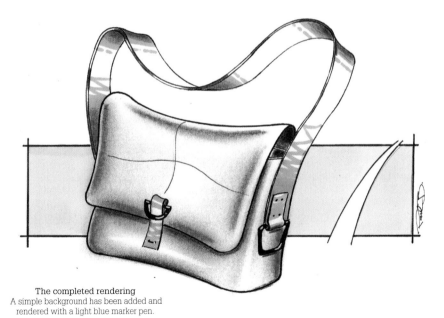

The completed rendering
A simple background has been added and
rendered with a light blue marker pen.

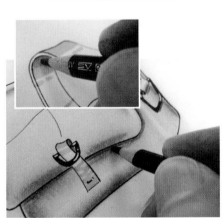

STEP 7
Heighten the shadow areas with a brown colouring pencil.

STEP 8
Outline the drawing with a 0.3 black fineline
pen against an ellipse template.

Electric kettle

Material: *shiny metal*

Media: *pastel pencils, marker pens, 0.3, 0.1 black fineline pens, white gel pen*

STEP 1
Redraw the outline of the kettle. (Use an ellipse template.)

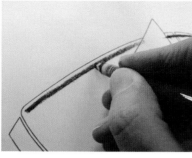

STEP 2
Apply pastel pencil to follow the curve of the side.

STEP 3
Wrap your finger in a tissue and smudge the pastel with your finger.

STEP 4
Use an eraser against an ellipse template to make a white edge.

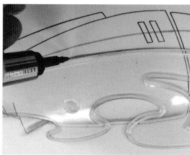

STEP 5
Use marker pen to define the dark areas.

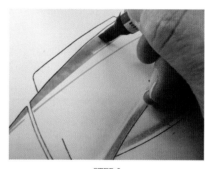

STEP 6
Apply marker to the appropriate areas.

STEP 7
Apply pastel pencil to the edge of the marker surface, then smudge into the kettle with your finger. (The colour must match the marker pen used.)

The completed rendering
A background and power lead have been added for visual interest.

STEP 8
Use the sharp edge of an eraser against an ellipse template to make the shiny stripes.

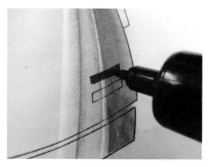

STEP 9
Render the vents with a black marker pen. Add highlights to the bottom edges with a white gel pen (see finished drawing).

STEP 10
Use an ellipse template to outline the drawing with a black fineline pen.

ISBN: 9780170355575

Drinking glass

Material: *glass*

Media: *pastel pencils, blue marker pen, 0.1 black fineline pen, white gel pen, light blue colouring pencil*

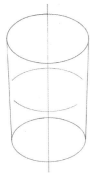

STEP 1
On craft paper, use an ellipse template to draw the glass. (Larger at the top, smaller at the bottom.)

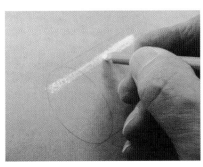

STEP 2
Apply white pastel pencil stripes that follow the angle of the side.

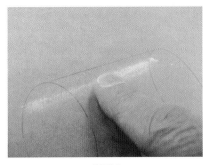

STEP 3
Smudge the pastel with your finger. Follow the angle of the side.

STEP 4
Use an ellipse template and a 0.1 black fineline pen to outline the ellipses. (Add a double line for the thickness of the glass at the top and bottom.)

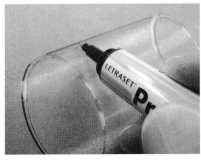

STEP 5
Apply a stripe of light blue marker down the sides of the glass.

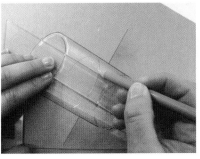

STEP 6
Apply a sharp, thin stripe of light blue colouring pencil on one side of the pastel. (Use the edge of a set square.)

STEP 7
Draw a thin black curvy line through the base of the glass and add sparkle with a white gel pen.

STEP 8
Add sparkle to the top edge with a white gel pen and a drop of white gouache.

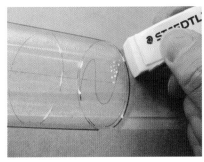

STEP 9
For reflection, smudge the white pastel sides into the craft paper beneath the glass. For added realism, follow the curve of the bottom of the glass with your eraser to erase a thin stripe.

The completed rendering
A horizon line with 4B pencil smudged below it anchors the drawing to the paper.

SITUATION

In your senior years at college you need a work space of your own at home.

BRIEF: Design a homework workstation.
DURATION: 20 weeks (class time)

REQUIREMENTS

1   Design research. *Cut and pasted among your sketches.*
2   Ideation of the workstation. Your sketches should show five stages: *thumbnails, exploratives, alternatives, thinking, and technical sketches. Consider human factors. Indicate your chosen design.* Four A3 pages minimum.
3   Place notes at any stage, about your research and sketches, to explain your thinking and details that are not clear visually.
4   Demonstrate in your sketches the effective development of your design ideas, *e.g. explore alternative ideas, not limited to only the brief and its requirements, then refine your chosen ideas informed by aesthetics and function. Support design judgements with qualitative and/or quantitative data gained through research, reflecting your values, tastes and/or views. Where appropriate, use drawings, models, digital modelling, etc.*

5   Draw with instruments, a detailed, scaled third angle orthographic projection of your workstation design. Show the following:
   - A minimum of three views, fully dimensioned, with one elevation sectioned.
   - Cutting planes, hidden detail, cross-hatching, notation and reference lines.
   - A title block with the scale and projection symbol.
   - Correct drawing standards and conventions.
6   Draw with instruments, to scale, the following:
   - a paraline drawing, in either oblique or isometric, of your chosen design
   - exploded isometric drawings of the main parts of the design to show construction detail.
7   Using media of your choice, produce a hand-rendered presentation drawing of your design that effectively communicates its shape and surface qualities.

DESIGN SPECIFICATIONS

The design must have curved or angled parts.
- Consideration of both aesthetics and function.
   Function: anthropometrics (chair and work surface height and sizes, etc.).
   Aesthetics: style, proportion, finish, harmony, form, etc.

Ideation

THUMBNAILS AND EXPLORATIVE SKETCHES

Here is an example of a product design brief at Curriculum Level 6. The intention at this level is to develop in the student a higher degree of sketching and thinking skills, a wider range of learning experiences, and important assessment opportunities.

Freehand sketching techniques need to be developed to a higher level of sophistication. Careful attention must be given to ensuring a mix of 2D and 3D sketches, orderly page layout, a range of viewpoints, a colour palette for the page (cool or hot colours), and a page background.

At the same time, plenty of white space should be left about the sketches to allow them to 'breathe'.

Colour palette has been considered carefully by all students.

Rendering is intentionally minimalist to provide adequate information about form and surface texture. Shadows beneath the sketches provide depth while backgrounds ground the work to the page.

Good use of black ballpoint enhances the thumbnails. Research also features on the pages to inform decisions, while notes explain thinking not clear visually.

Effective use of craft paper gives the pages a 'work in progress' feel. (Using a paper other than white provides a 'non-scary' background and aids in the quick thinking of ideas.)

Ballpoint pens force the flow of ideas without having to think there is a mistake to erase.

Although some handwritten titles have been done, the pages were placed into an A3 printer and the titles printed out from a computer before drawing began.

SKILLS NEEDED TO PRECEDE THIS PAGE:

Planning sheet.
Freehand design sketching — a mix of 2D and 3D, varying viewpoint, backgrounds.
Crating (proportions) and line hierarchy.
Rendering techniques — light source, tonal changes, form.
Research and simple design notes that explain thinking not clear visually.
Backgrounds and shadows.
Media technique: marker pens, black fineline pens, pastel pencils.
Links to: pages 75–81

Nathan Yee

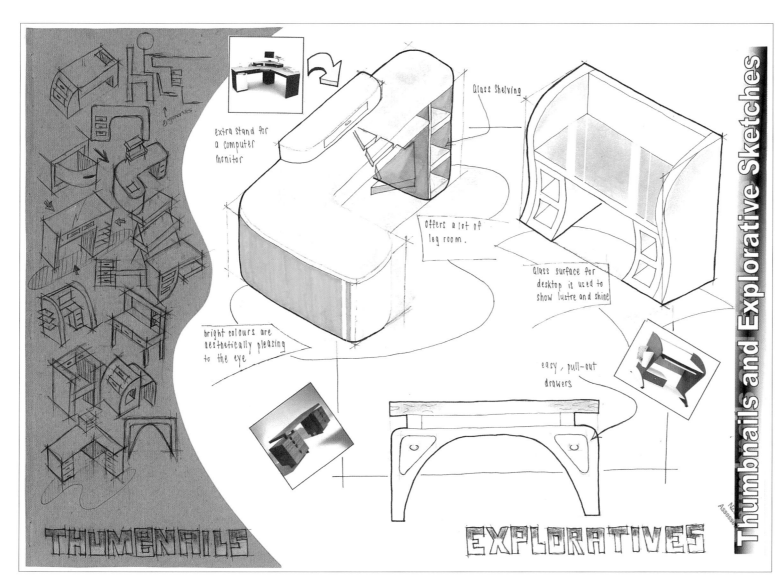

extra stand for a computer monitor

Glass Shelving

Offers a lot of leg room.

Glass surface for desktop is used to show lustre and shine

bright colours are aesthetically pleasing to the eye

easy, pull-out drawers

THUMBNAILS

EXPLORATIVES

I thought a lot about the layout of this page before I started. I wanted it to be pleasing and allow the eye to follow each design I explored.

I did a lot of research. If it wasn't for this, I would have struggled to explore my thumbnails further. They formed the basis of my design ideas.

I kept the rendering deliberately simple, using marker pens and pastel pencil to show materials like wood and shiny surfaces.

I used black fineline pens to outline the sketches. The thick and thin lines and shadows made my designs pop out from the page.

Belinda Van Eeden

I found the thumbnails difficult to do because I like to be neat and thought they might be messy. Once I realised that the idea is to work fast and not worry too much about how they look, I found my ideas flowed easier. Using craft paper and a ballpoint pen forced me not to think about making mistakes.

I planned the page layout first to make sure it looked effective.

For the sketches I deliberately used steeper angles on the crates and explored different viewpoints to better show off my thinking. I like the way the thick and thin outlines of the sketches look.

I was a bit worried about how to render the drawings at first, but decided to keep it simple by shading the parts furthest from the light. I used marker pen and pastel and added a shadow for effect. I think my squiggled lines work well to show materials and shiny surfaces. The background is smudged pastel and helps to tie the design together.

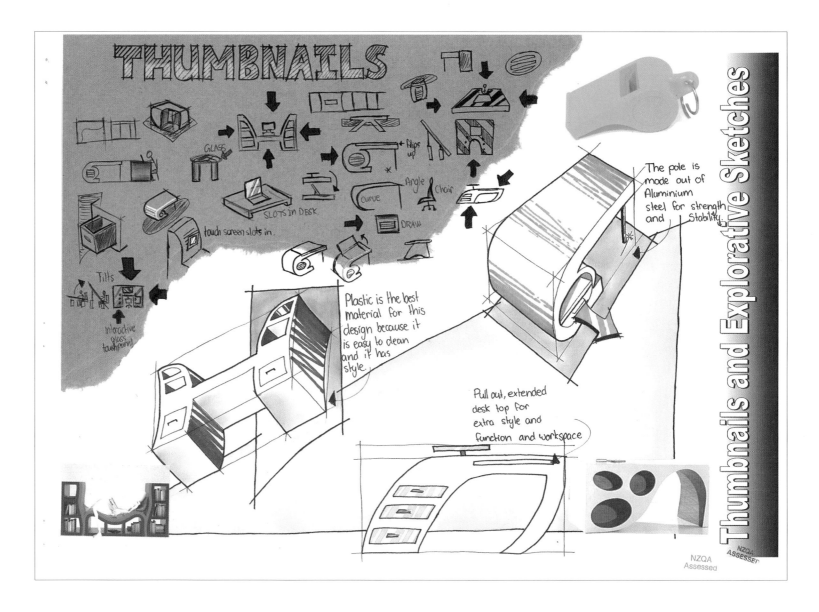

ISBN: 9780170355575

Stephan Gailer

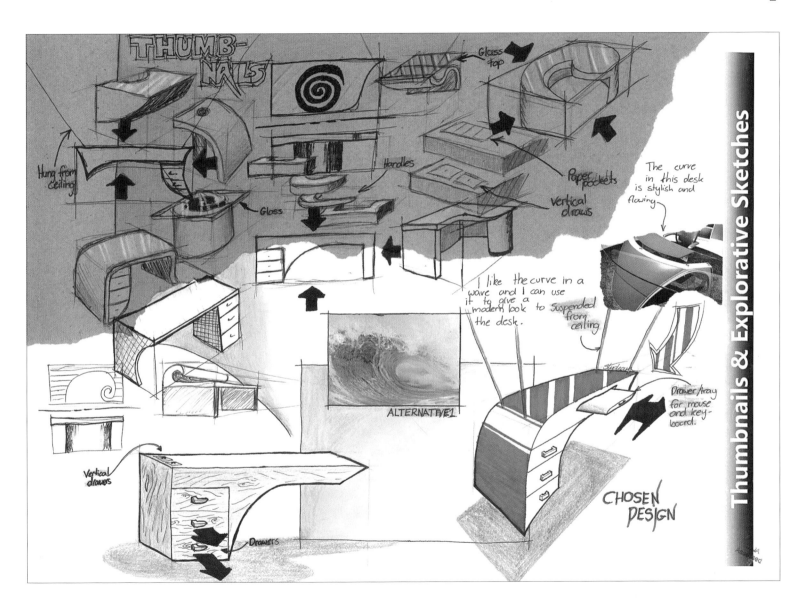

Thumbnails & Explorative Sketches

It took me a long time to think about the design I wanted.

From the start I thought about a design that would be different. When I did the thumbnails, I found an idea of maybe having a design that was suspended from the ceiling.

I found some good research that helped with getting the curved parts.

Even though my chosen design is on this page, if I was to do it again I would explore more design ideas.

I planned the page layout on a planning sheet first. My designs are long so that's why I put the thumbnails across the top to give better space for the exploratives.

I thought about the colours for the page, and used orange pastel for the backgrounds. Colouring pencil and marker pen was used for rendering the designs.

I practised drawing on steeper angles first, before I did the chosen design sketch.

SITUATION
In your senior years at college you need a work space of your own at home.

BRIEF: Design a homework workstation.
DURATION: 20 weeks (class time)

REQUIREMENTS
1   Design research. *Cut and pasted among your sketches.*
2   Ideation of the workstation. Your sketches should show five stages: *thumbnails, exploratives, alternatives, thinking, and technical sketches. Consider human factors. Indicate your chosen design.* Four A3 pages minimum.
3   Place notes at any stage, about your research and sketches, to explain your thinking and details that are not clear visually.
4   Demonstrate in your sketches the effective development of your design ideas, *e.g. explore alternative ideas, not limited to only the brief and its requirements, then refine your chosen ideas informed by aesthetics and function. Support design judgements with qualitative and/or quantitative data gained through research, reflecting your values, tastes and/or views. Where appropriate, use drawings, models, digital modelling, etc.*

5   Draw with instruments, a detailed, scaled third angle orthographic projection of your workstation design. Show the following:
    ■   A minimum of three views, fully dimensioned, with one elevation sectioned.
    ■   Cutting planes, hidden detail, cross-hatching, notation and reference lines.
    ■   A title block with the scale and projection symbol.
    ■   Correct drawing standards and conventions.
6   Draw with instruments, to scale, the following:
    ■   a paraline drawing, in either oblique or isometric, of your chosen design
    ■   exploded isometric drawings of the main parts of the design to show construction detail.
7   Using media of your choice, produce a hand-rendered presentation drawing of your design that effectively communicates its shape and surface qualities.

DESIGN SPECIFICATIONS
The design must have curved or angled parts.
■   Consideration of both aesthetics and function.
    Function: anthropometrics (chair and work surface height and sizes, etc.).
    Aesthetics: style, proportion, finish, harmony, form, etc.

Ideation (cont.)

ALTERNATIVE SKETCHES

After initial exploration of ideas, students must be given the opportunity to extend their thinking beyond just the brief and its requirements.

Exploration of alternative ideas will better inform the design, resulting in a more resolved final solution. Students need to be encouraged to look elsewhere for inspiration and to think 'outside the square'.

Organic shapes provided by nature, as seen in all three students' work on the following pages, clearly have an influence on the design.

A lesson on bio-mimickry provokes thinking in the right direction.

Note the continuation of the same sketching style in 2D and 3D, and the dramatic use of viewpoint to highlight shape and form. Particularly effective is Nathan's view with the sliding top.

Note also arrows that show function and/or movement, and the carefully considered colour palette.

Crating is clearly defined, offering correct proportion, while thick and thin lines (line hierarchy) make the sketches 'pop' on the page.

Sketching on craft paper adds to the 'work in progess' feel to the pages.

Once again, minimalist marker rendering conveys tone and texture effectively.

White space, a background that draws the eye to the flow of ideas, and shadows enhance the page.

SKILLS NEEDED TO PRECEDE THIS PAGE:

Planning sheet used. Freehand design sketching in 2D and 3D.

Crating (proportions) and line hierarchy.

Research is ongoing to inform ideas.

Viewpoints are considered for visual interest and to better show the ideas.

Arrows indicate movement and/or function.

Rendering techniques — minimalist, indicating the light source, tonal changes and form.

Colour palette considered.

Links to: pages 75–81

Nathan Yee

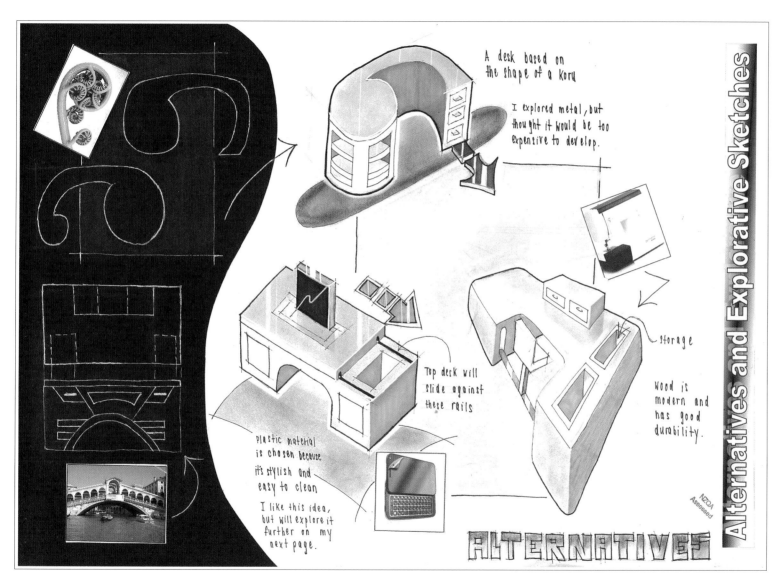

A desk based on the shape of a koru

I explored metal, but thought it would be too expensive to develop.

Top desk will slide against these rails

plastic material is chosen because it's stylish and easy to clean

I like this idea, but will explore it further on my next page.

storage

Wood is modern and has good durability.

NZQA Assessed

ALTERNATIVES

To make my design better, I looked at some shapes from nature and used this as research. This allowed me to take my design to the next stage.

To emphasise my research, I used black paper to add contrast to the page.

I used more steeper angles on these sketches because I wanted to show the top of the desks. I was worried at first that they wouldn't look right, but I think they turned out well.

Using thick and thin lines again, and some arrows to show how parts work, make the design easier to understand. But if I had to do the page again, I would improve the shadows so that they work better with the designs.

Belinda Van Eeden

I wasn't sure what to show on this page at first, so I researched some outdoor furniture with interesting shapes. These helped me come up with a better design.

I think the use of black craft paper and white pen for drawing with has made the page look much more effective.

I used some 3D and 2D sketches on this page, making the crate for my chosen design looking up from underneath to make it stand out from the others on the page.

I purposefully thought about how to render the sketches, and what colours to use, and decided again that keeping it simple works best. I used smudged pastel and some marker pen to shade the surfaces furthest from the light.

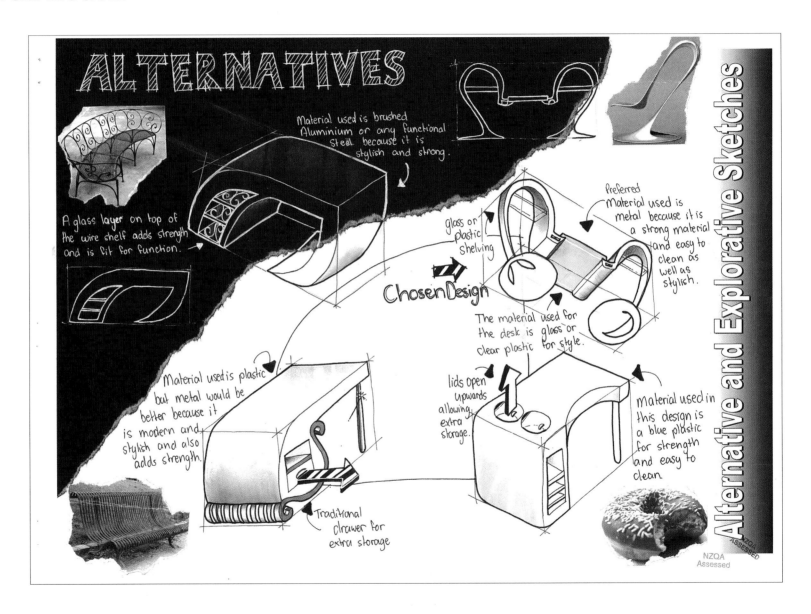

ALTERNATIVES

Material used is brushed Aluminium or any functional Steel because it is stylish and strong.

A glass layer on top of the wire shelf adds strength and is fit for function.

glass or plastic shelving

Preferred material used is metal because it is a strong material and easy to clean as well as stylish.

Chosen Design

The material used for the desk is glass or clear plastic for style.

Material used is plastic but metal would be better because it is modern and stylish and also adds strength.

lids open upwards allowing extra storage.

Material used in this design is a blue plastic for strength and easy to clean.

Traditional drawer for extra storage

Alternative and Explorative Sketches

ISBN: 9780170355575

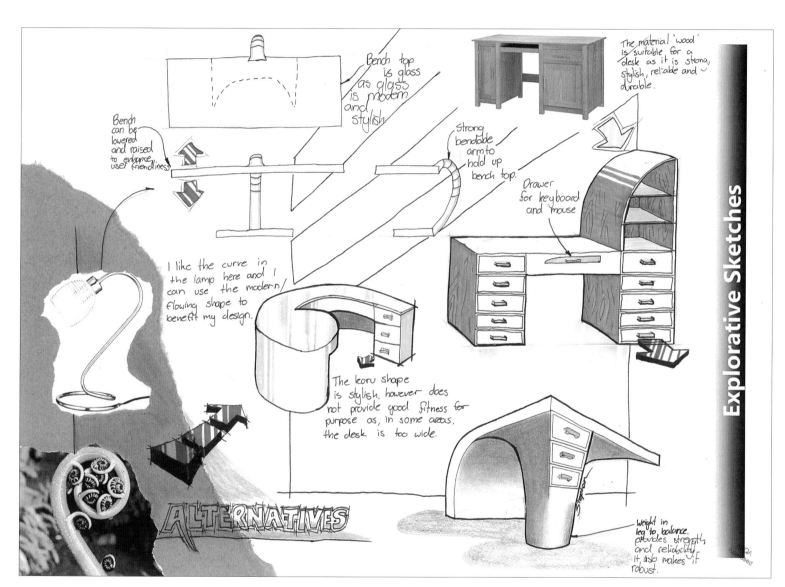

Bench top is glass as glass is modern and stylish

Bench can be lowered and raised to enhance user friendliness

The material 'wood' is suitable for a desk as it is strong, stylish, reliable and durable.

Strong bendable arm to hold up bench top.

Drawer for keyboard and mouse

I like the curve in the lamp here and I can use the modern/flowing shape to benefit my design.

The koru shape is stylish, however does not provide good fitness for purpose as, in some areas, the desk is too wide.

Explorative Sketches

ALTERNATIVES

Weight in leg to balance. provides strength and reliability. It also makes it robust.

Because I needed to show curves in the design, I concentrated on finding alternative research images with these in them.

Even though I found doing this page useful, and I like some of the ideas, I went back to my explorative page for my chosen design.

I used isometric, oblique and perspective sketches on this page. I like the way the bottom sketch turned out with the view looking up from below. 2D was easier for showing the idea based on the curved light.

I kept the rendering simple and only used marker pens and colouring pencils. The background is smudged pastel pencil.

I kept the same colour palette for this page as my first page, so that they would match each other.

Apart from the alternative title, I printed the title on the computer first, then did the drawings.

SITUATION
In your senior years at college you need a work space of your own at home.

BRIEF: Design a homework workstation.
DURATION: 20 weeks (class time)

REQUIREMENTS

1   Design research. *Cut and pasted among your sketches.*
2   Ideation of the workstation. Your sketches should show five stages: *thumbnails, exploratives, alternatives, thinking, and technical sketches. Consider human factors. Indicate your chosen design.* Four A3 pages minimum.
3   Place notes at any stage, about your research and sketches, to explain your thinking and details that are not clear visually.
4   Demonstrate in your sketches the effective development of your design ideas, *e.g. explore alternative ideas, not limited to only the brief and its requirements, then refine your chosen ideas informed by aesthetics and function. Support design judgements with qualitative and/or quantitative data gained through research, reflecting your values, tastes and/or views. Where appropriate, use drawings, models, digital modelling, etc.*

5   Draw with instruments, a detailed, scaled third angle orthographic projection of your workstation design. Show the following:
  ▪ A minimum of three views, fully dimensioned, with one elevation sectioned.
  ▪ Cutting planes, hidden detail, cross-hatching, notation and reference lines.
  ▪ A title block with the scale and projection symbol.
  ▪ Correct drawing standards and conventions.
6   Draw with instruments, to scale, the following:
  ▪ a paraline drawing, in either oblique or isometric, of your chosen design
  ▪ exploded isometric drawings of the main parts of the design to show construction detail.
7   Using media of your choice, produce a hand-rendered presentation drawing of your design that effectively communicates its shape and surface qualities.

DESIGN SPECIFICATIONS
The design must have curved or angled parts.
  ▪ Consideration of both aesthetics and function.
    Function: anthropometrics (chair and work surface height and sizes, etc.).
    Aesthetics: style, proportion, finish, harmony, form, etc.

Ideation (cont.)

THINKING SKETCHES AND HUMAN FACTORS

Here is a most important stage in the ideation process, particularly for product design.

Now is the time to think about what the design could be made from, and to 'fine tune' it to fit the human form.

These pages show the students' exploration of the human form, materials that could be used in their workstation designs, the sizes of the design (important for the scaled instrumental drawings to follow) and other detail such as a hand grip for the drawers.

A chosen design decision should be getting close to being made around this stage. A good idea is to indicate this on the page as Nathan has done.

Note the continued use of crating for correct proportions, 2D and 3D sketches, viewpoints and line hierarchy (thick and thin lines) that provide the sketches with visual impact.

Research is also ongoing, continuously informing ideas. Simple notes explain thinking that is not clear visually.

Again, rendering indicates tone, surface shape and texture. Careful consideration of a colour palette enhances the sketches. Particularly effective is Stephan's use of blacks and blues.

The use of craft paper continues to provide visual interest to the page and that 'work in progress' feel. Lots of white space allows the sketches to 'breathe' on the page. Backgrounds are effective in grouping sets of thinking together.

SKILLS NEEDED TO PRECEDE THIS PAGE:

Planning sheet used.
Human factors and sizes indicated. (Required for any instrumental drawings to follow.)
Ongoing research better informs decisions.
Freehand design sketching in 2D and 3D with clearly defined rating and line hierarchy.
Viewpoints are considered for visual interest and to better show the ideas.
Rendering techniques — minimalist, indicating the light source, tonal changes, shape and surface texture.
Colour palette considered.
Links to: pages 75–81

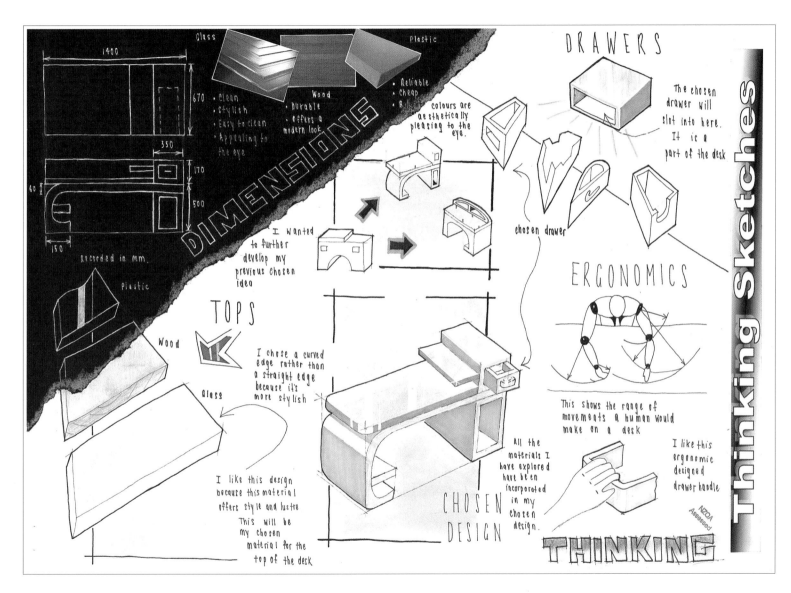

DIMENSIONS

1400

670

350

170

60

500

150

Recorded in mm.

Plastic

Glass
- Clean
- Stylish
- Easy to clean
- Appealing to the eye

Wood
- Durable
- Offers a modern look

Plastic
- Reliable
- Cheap
- R...

colours are aesthetically pleasing to the eye.

TOPS

Wood

Glass

I chose a curved edge rather than a straight edge because it's more stylish

I like this design because this material offers style and lustre. This will be my chosen material for the top of the desk

I wanted to further develop my previous chosen idea

DRAWERS

The chosen drawer will slot into here. It is a part of the desk

chosen drawer

ERGONOMICS

This shows the range of movements a human would make on a desk

I like this ergonomic designed drawer handle

CHOSEN DESIGN

All the materials I have explored have been incorporated in my chosen design.

THINKING

NZQA Assessed

Thinking Sketches

This is my favourite page. I think it is set out well and makes my design easy to understand.

I did quite a bit of research into materials, sizes and shapes. I also looked at lots of different drawer types to get the one I liked. By starting to think about human factors, I got an idea for how wide the desk should be. All these things allowed me to come up with my chosen design.

I was careful when drawing the crate for this because it was important to make sure it was the right proportions and realistic to my final design.

I used simple rendering again, using mainly marker pen and smudged pastel for the background.

Belinda Van Eeden

At this stage of my design, I needed to think about the sizes more. I considered how it could fit the human form ... so I did some drawings to show the measurement of the human arm span and leg space.

I did some research into materials to show what my design could be made from. This also helped me when I rendered my drawings as I could use colours to represent the materials. That's why the page is a bit more multi-coloured than my others.

The bit I found the hardest was how to find a stylish way to show all these details, so that all my pages continued to match each other in layout and look.

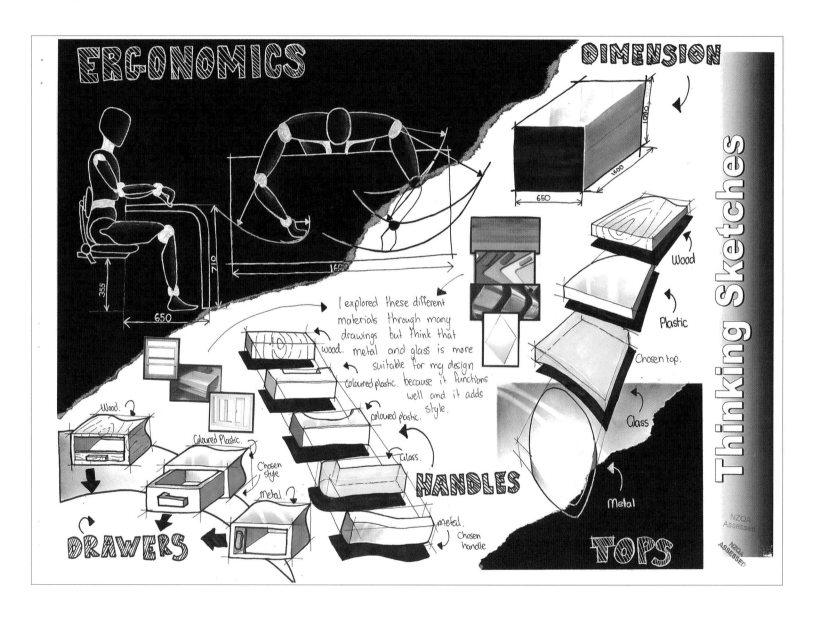

ISBN: 9780170355575

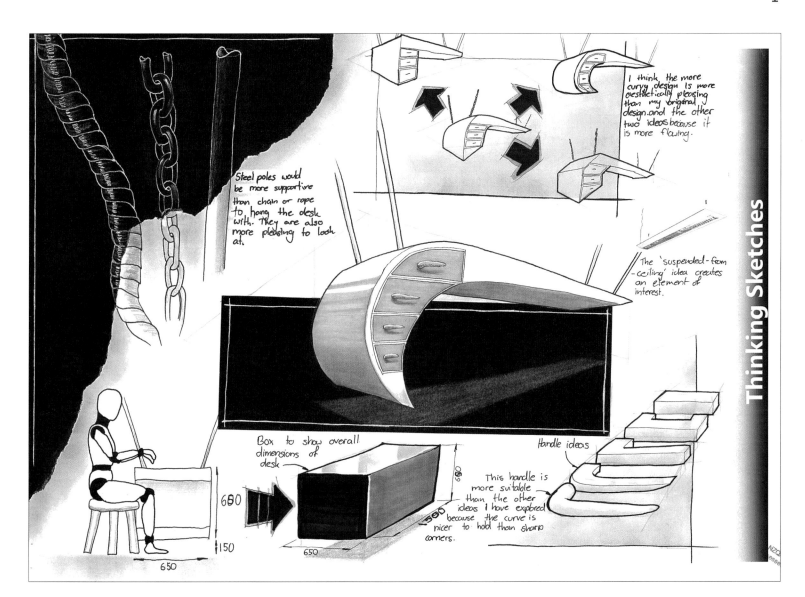

Thinking Sketches

I think the more curvy design is more aesthetically pleasing than my original design, and the other two ideas because it is more flowing.

Steel poles would be more supportive than chain or rope to hang the desk with. They are also more pleasing to look at.

The 'suspended-from-ceiling' idea creates an element of interest.

Box to show overall dimensions of desk

650

680

150

650

Handle ideas

This handle is more suitable than the other ideas I have explored because the curve is nicer to hold than sharp corners.

At this stage I was sure about my chosen design. I went with the idea of hanging it from the roof.

I did some more sketches at the top of the page to get the final shape that I wanted.

The hardest part was deciding on what to use to hang the desk from the roof. I did lots of research and decided on using steel poles.

I planned the page layout first. I wanted my design to be clearly seen so put it in the middle of the page. If I was to do it again, I would have shown more details of how to attach the steel rods.

Getting the sizes was easy. I just measured my desk at home.

I wanted a different look to this page so I rendered the desk with blue markers. I used black paper for the background and cut and pasted it around the drawing.

I used a white pen and pastel pencil on black paper to show the chain and rope ideas.

SITUATION

In your senior years at college you need a work space of your own at home.

BRIEF: Design a homework workstation.
DURATION: 20 weeks (class time)

REQUIREMENTS

1. Design research. *Cut and pasted among your sketches.*
2. Ideation of the workstation. Your sketches should show five stages: *thumbnails, exploratives, alternatives, thinking, and technical sketches. Consider human factors. Indicate your chosen design.* Four A3 pages minimum.
3. Place notes at any stage, about your research and sketches, to explain your thinking and details that are not clear visually.
4. Demonstrate in your sketches the effective development of your design ideas, *e.g. explore alternative ideas, not limited to only the brief and its requirements, then refine your chosen ideas informed by aesthetics and function. Support design judgements with qualitative and/or quantitative data gained through research, reflecting your values, tastes and/or views. Where appropriate, use drawings, models, digital modelling, etc.*

5. Draw with instruments, a detailed, scaled third angle orthographic projection of your workstation design. Show the following:
   - A minimum of three views, fully dimensioned, with one elevation sectioned.
   - Cutting planes, hidden detail, cross-hatching, notation and reference lines.
   - A title block with the scale and projection symbol.
   - Correct drawing standards and conventions.
6. Draw with instruments, to scale, the following:
   - a paraline drawing, in either oblique or isometric, of your chosen design
   - exploded isometric drawings of the main parts of the design to show construction detail.
7. Using media of your choice, produce a hand-rendered presentation drawing of your design that effectively communicates its shape and surface qualities.

DESIGN SPECIFICATIONS

The design must have curved or angled parts.
- Consideration of both aesthetics and function.
  Function: anthropometrics (chair and work surface height and sizes, etc.).
  Aesthetics: style, proportion, finish, harmony, form, etc.

Ideation (cont.)

TECHNICAL SKETCHES

Once a final design has been chosen, it is time to think about the construction details of the design that could allow it to be made.

Technical sketches need to be a mix of 3D exploded isometric and 2D sectioned views of parts. They do not need to be the entire design, only parts that are important to the manufacture of the design.

All students have shown this well, at the same time confirming their understanding not only of how the design could be made, but also their knowledge of sketching techniques.

Note the size of each sketch. Larger sketches make understanding clearer, and allow students to show construction details of the sketch.

Note also that the parts are drawn pulled apart in the direction in which they will be assembled.

Good use of line hierarchy (thick and thin lines) provides the sketches with visual impact. Rendering is minimalist to highlight tone and surface shape.

Backgrounds and colour palette continue the theme from earlier pages.

The use of craft paper provides visual interest. This is also a useful technique that can be employed to cover up any errors that may cause the student to start the page again.

SKILLS NEEDED TO PRECEDE THIS PAGE:

Planning sheet used. Freehand design sketching in 2D and 3D. Crating and line hierarchy.

Construction techniques resource books and internet.

Viewpoints are considered for visual interest and to better show the ideas.

Correct 'exploded' techniques. (The parts are shown pulled apart in the direction in which they would be assembled.)

Correct 2D cross-hatching technique from a cutting plane.

Backgrounds and colour palette considered.

Simple notes explain thinking not clear visually.

Rendering techniques — minimalist, indicating the light source, tonal changes and surface texture.

Links to: pages 75–81, 187

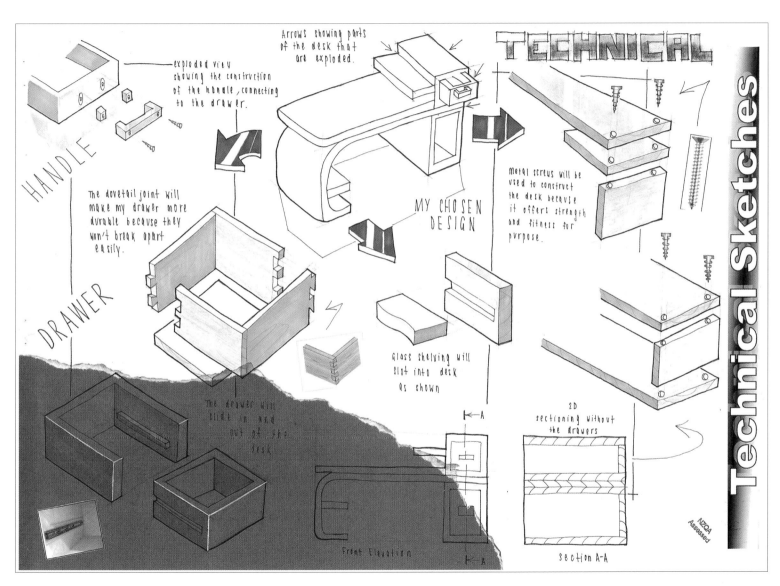

TECHNICAL

exploded view showing the construction of the handle, connecting to the drawer.

Arrows showing parts of the desk that are exploded.

HANDLE

DRAWER

The dovetail joint will make my drawer more durable because they won't break apart easily.

MY CHOSEN DESIGN

metal screws will be used to construct the desk because it offers strength and fitness for purpose.

Glass shelving will slot into desk as shown

The drawer will slide in and out of the desk

Front Elevation

2D sectioning without the drawers

Section A-A

A

Technical Sketches

This was the most challenging of all the pages for me. I needed to find out how my design could be made, and show this in sketches.

Research again was a big help. I tried to make sure that the things I was showing would actually work. I wanted to make sure it would be durable and have a good finish.

I used steeper angles on the 3D sketches again on this page, to make them stand out and to show details better.

I did a 2D sectioned sketch to show the inside of the desk. This also gave me some ideas that I would use later in the orthographic projection drawing.

NZQA Assessed

Belinda Van Eeden

I found this page quite hard. I had to research existing designs to give me ideas of how to go about constructing mine.

The most difficult part was trying to show people reading my drawings, how my design would move, work and function.

I found that the use of bold arrows made my ideas easier to understand the way the parts would join together.

For the sectioned view, I was glad that I had done some sectioned orthographic drawings in class beforehand. By using this type of drawing I think my design is communicated much clearer.

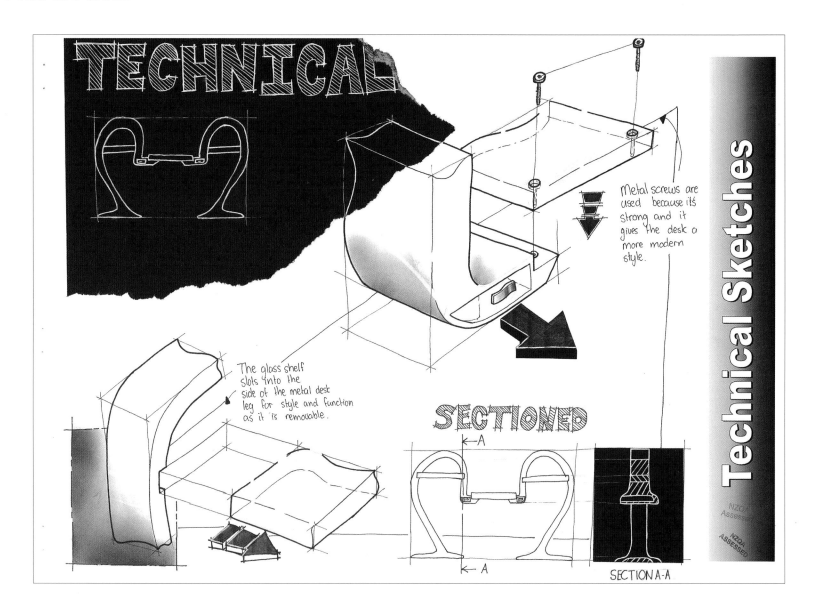

TECHNICAL

Metal screws are used because it's strong and it gives the desk a more modern style.

The glass shelf slots into the side of the metal desk leg for style and function as it is removable.

SECTIONED

←—A

←— A

SECTION A-A

Technical Sketches

ISBN: 9780170355575

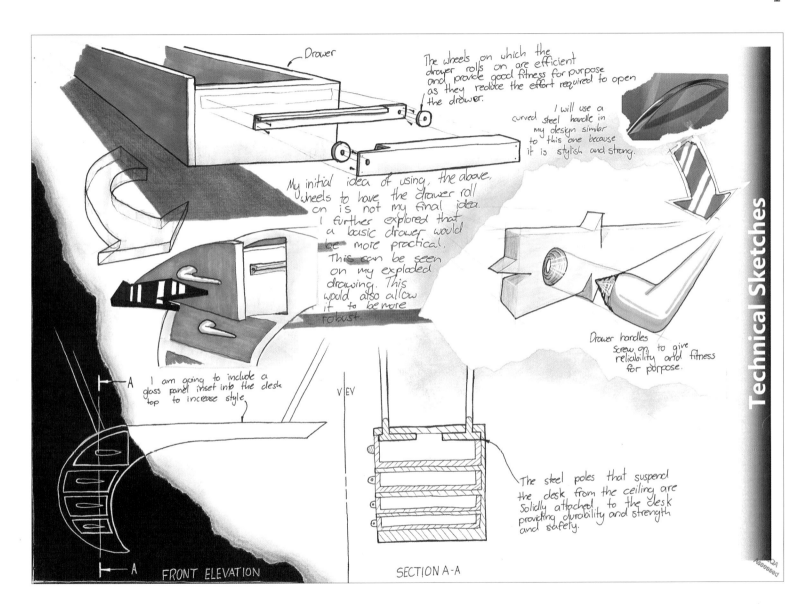

Drawer

The wheels on which the drawer rolls on are efficient and provide good fitness for purpose as they reduce the effort required to open the drawer.

I will use a curved steel handle in my design similar to this one because it is stylish and strong.

My initial idea of using, the above, wheels to have the drawer roll on is not my final idea. I further explored that a basic drawer would be more practical. This can be seen on my exploded drawing. This would also allow it to be more robust.

A — I am going to include a glass panel inset into the desk top to increase style.

VIEW

Drawer handles screw on, to give reliability and fitness for purpose.

The steel poles that suspend the desk from the ceiling are solidly attached to the desk providing durability and strength and safety.

FRONT ELEVATION

SECTION A-A

Technical Sketches

For these technical sketches, I concentrated on showing how the drawer could be made, and how the desk could hang from the ceiling.

I did a lot of research beforehand to find out how these parts could be made.

I like the way the exploded drawing shows the drawer clearly.

It was hard to show the steel rods in a 3D sketch, so I used a 2D sectioned view.

The exploded view of the drawer handle was done on a separate piece of paper, then pasted over the top of another sketch, which I did not like. I found this to be a good way to cover up mistakes and still keep the page looking good.

I used the same colours and rendering media as the previous page. I like the way it turned out and the consistency it provided within my design pages.

SITUATION
In your senior years at college you need a work space of your own at home.

BRIEF: Design a homework workstation.
DURATION: 20 weeks (class time)

REQUIREMENTS
1   Design research. *Cut and pasted among your sketches.*
2   Ideation of the workstation. Your sketches should show five stages: *thumbnails, exploratives, alternatives, thinking, and technical sketches. Consider human factors. Indicate your chosen design.* Four A3 pages minimum.
3   Place notes at any stage, about your research and sketches, to explain your thinking and details that are not clear visually.
4   Demonstrate in your sketches the effective development of your design ideas, *e.g. explore alternative ideas, not limited to only the brief and its requirements, then refine your chosen ideas informed by aesthetics and function. Support design judgements with qualitative and/or quantitative data gained through research, reflecting your values, tastes and/or views. Where appropriate, use drawings, models, digital modelling, etc.*

5   Draw with instruments, a detailed, scaled third angle orthographic projection of your workstation design. Show the following:
    - A minimum of three views, fully dimensioned, with one elevation sectioned.
    - Cutting planes, hidden detail, cross-hatching, notation and reference lines.
    - A title block with the scale and projection symbol.
    - Correct drawing standards and conventions.

6   Draw with instruments, to scale, the following:
    - a paraline drawing, in either oblique or isometric, of your chosen design
    - exploded isometric drawings of the main parts of the design to show construction detail.
7   Using media of your choice, produce a hand-rendered presentation drawing of your design that effectively communicates its shape and surface qualities.

DESIGN SPECIFICATIONS
The design must have curved or angled parts.
- Consideration of both aesthetics and function.
  Function: anthropometrics (chair and work surface height and sizes, etc.).
  Aesthetics: style, proportion, finish, harmony, form, etc.

Instrumental drawing

THIRD ANGLE ORTHOGRAPHIC PROJECTION

Instrumental drawings are often overlooked. But this is a shame.

As seen in all three drawings, they offer important, accurate, scaled information on sizes and construction that could allow the design to be manufactured.

Instrumental drawings also show another aspect of student drawing ability, complementing the freedom of ideation pages.

Before these drawings can be attempted, comprehensive classwork exercises must be undertaken, directed at every step by the teacher.

Third angle orthographic projection needs to be introduced in earlier years of study, so that students are not attempting to draw their designs in this format from the outset without any prior knowledge.

This drawing method shows a wide range of essential drawing skills that include dimensioning, cutting planes, cross-hatching and a parts list.

Hint: Take a photocopy of the part that contains a curved or angled surface, so that it can be used as an auxiliary view on the paraline drawing page. This will facilitate plotting of that part into isometric or oblique (see next set of drawings).

SKILLS NEEDED TO PRECEDE THIS PAGE:

Setting out the drawing to fit the page, line weights and line standards (2H pencil for all lines).
Title block, scales and projection symbol.
Printing standards — 5 mm in title block, 3 mm about the drawing (HB pencil).
Correct dimensioning techniques for third angle projection.
Cutting plane and cross-hatching, hidden detail, reference lines, notation.
A parts list with numbered components about the views.
Views labelled.
Links to: pages 71, 184, 185

ISBN: 9780170355575   PHOTOCOPYING OF THIS PAGE IS RESTRICTED UNDER LAW.

Nathan Yee

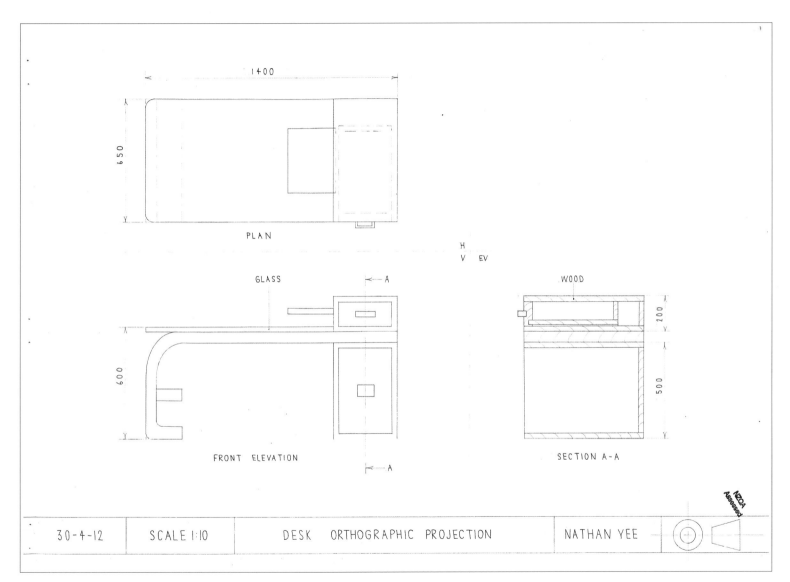

PLAN

1400

650

H
V   EV

FRONT ELEVATION

GLASS

A

600

A

SECTION A-A

WOOD

200

500

| 30-4-12 | SCALE 1:10 | DESK   ORTHOGRAPHIC   PROJECTION | NATHAN YEE | |

By having previous knowledge of orthographic drawings, and after having done lots of them since starting this subject, this drawing was not as hard as I thought it would be.

Getting the scale right, and setting out the drawing on the page, took a while. Hidden detail, sectioning and the cutting plane detail also took a lot of thinking. I kept having to make adjustments to make sure the drawing was as accurate as possible.

Making sure I used three different line types at the right time on different parts of the drawing was a challenge. I used a clutch pencil with a 2H lead in it. I'm glad I took time to get all the lines right, as it did add a much neater and professional look.

Belinda Van Eeden

This drawing was a challenge. It took some time to work out the right scale and to get a layout that fitted the page. At the same time I also needed to make sure I showed enough details of how it could be made.

Once I had all the details sorted out, I found the actual drawing did not take as long as I thought it would.

I had to keep making sure that I was using the correct pencil weights, and getting all the dimensions right, because I knew this drawing was going to be marked for my assessment.

Having done these types of drawings in class beforehand, and knowing how to make a title block, really helped a lot. I would recommend to anyone that a practice drawing is done first.

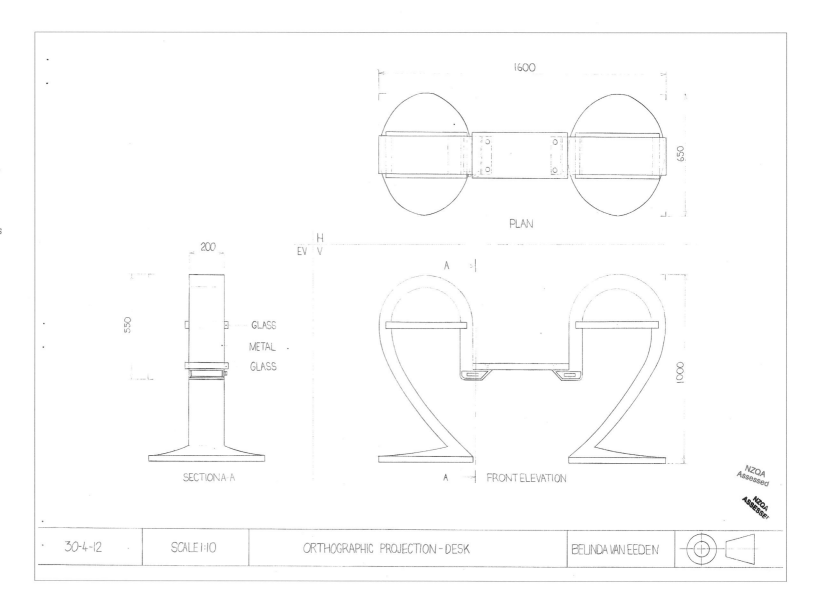

Stephan Gailer

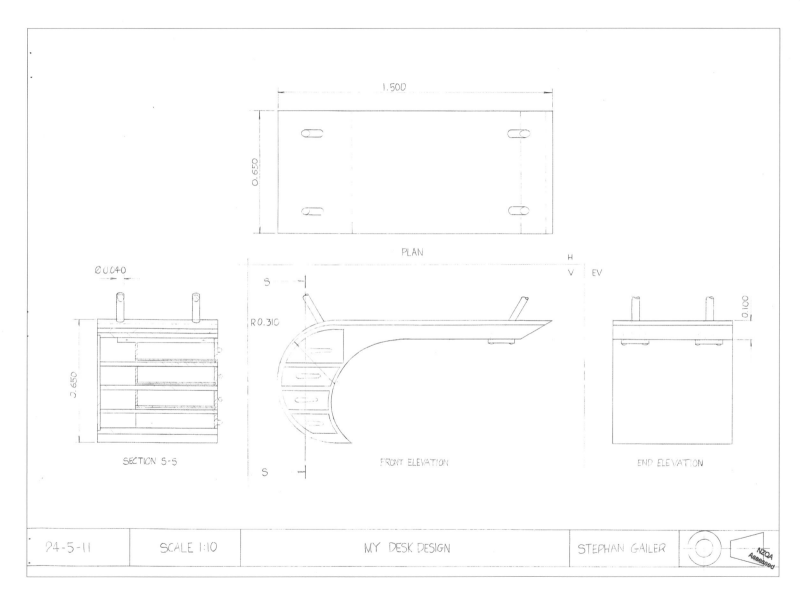

1.500

0.650

PLAN

Ø 0.040

R 0.310

0.650

0.100

SECTION S-S

FRONT ELEVATION

END ELEVATION

H
V   EV
S

S

24-5-11	SCALE 1:10	MY DESK DESIGN	STEPHAN GAILER	NZQA Assessed

This drawing took a lot of time. Because the desk is large, it took a while to get the right scale. I was going to show only one end elevation, but wanted to show two for more detail.

I did a practice page first and was glad I had done lots of these types of drawings in class beforehand.

I used a compass and an ellipse template for the curved parts in the front elevation.

The view that took the most time was the cross-section end elevation. I researched how drawers are made so I could show them accurately.

I did the drawing in construction lines first, then spent a lot of time checking it before I outlined it. I knew I had to get all the lines right to get good marks.

I used a clutch pencil with a 2H lead for the lines, and an HB lead for the title block printing, labels and dimensions.

I like the way it turned out, and am glad I spent the time to get it right.

SITUATION

In your senior years at college you need a work space of your own at home.

BRIEF: Design a homework workstation.
DURATION: 20 weeks (class time)

REQUIREMENTS

1   Design research. *Cut and pasted among your sketches.*
2   Ideation of the workstation. Your sketches should show five stages: *thumbnails, exploratives, alternatives, thinking, and technical sketches. Consider human factors. Indicate your chosen design.* Four A3 pages minimum.
3   Place notes at any stage, about your research and sketches, to explain your thinking and details that are not clear visually.
4   Demonstrate in your sketches the effective development of your design ideas, *e.g. explore alternative ideas, not limited to only the brief and its requirements, then refine your chosen ideas informed by aesthetics and function. Support design judgements with qualitative and/or quantitative data gained through research, reflecting your values, tastes and/or views. Where appropriate, use drawings, models, digital modelling, etc.*
5   Draw with instruments, a detailed, scaled third angle orthographic projection of your workstation design. Show the following:
    ▪ A minimum of three views, fully dimensioned, with one elevation sectioned.
    ▪ Cutting planes, hidden detail, cross-hatching, notation and reference lines.
    ▪ A title block with the scale and projection symbol.
    ▪ Correct drawing standards and conventions.

6   Draw with instruments, to scale, the following:
    ▪ a paraline drawing, in either oblique or isometric, of your chosen design

    ▪ exploded isometric drawings of the main parts of the design to show construction detail.
7   Using media of your choice, produce a hand-rendered presentation drawing of your design that effectively communicates its shape and surface qualities.

DESIGN SPECIFICATIONS

The design must have curved or angled parts.
▪ Consideration of both aesthetics and function.
    Function: anthropometrics (chair and work surface height and sizes, etc.).
    Aesthetics: style, proportion, finish, harmony, form, etc.

Instrumental drawing (cont.)

PARALINE DRAWING

Here is another instrumental drawing often overlooked.

To complement an orthographic projection, this drawing clarifies information that could allow the design to be manufactured. Again, it serves to showcase another aspect of student drawing ability and complement the freedom of ideation pages.

Hint: Asking for curved and/or angled parts to be incorporated into the design (see the specifications at the bottom of the brief) gives students opportunities to gain higher achievement at assessment time.

As mentioned earlier, before these drawings can be attempted, comprehensive classwork exercises must be undertaken, directed at every step by the teacher.

Paraline drawing (isometric and oblique) needs to be introduced in earlier years of study, so that students are not attempting to draw their designs without any prior knowledge.

The drawings featured show a wide range of essential instrumental drawing skills that include circle and/or curve constructions.

Note also the addition of an auxiliary view (in this case, a photocopy of the curved part from the orthographic projection), which facilitates the plotting of the curved part into isometric.

 SKILLS NEEDED TO PRECEDE THIS PAGE:

Line weights and line standards (2H pencil for all lines).
Title block and scales.
Printing standards — 5 mm in title block, 3 mm about the drawing (HB pencil).
3D circle and curve constructions.
Links to: pages 186–187

Nathan Yee

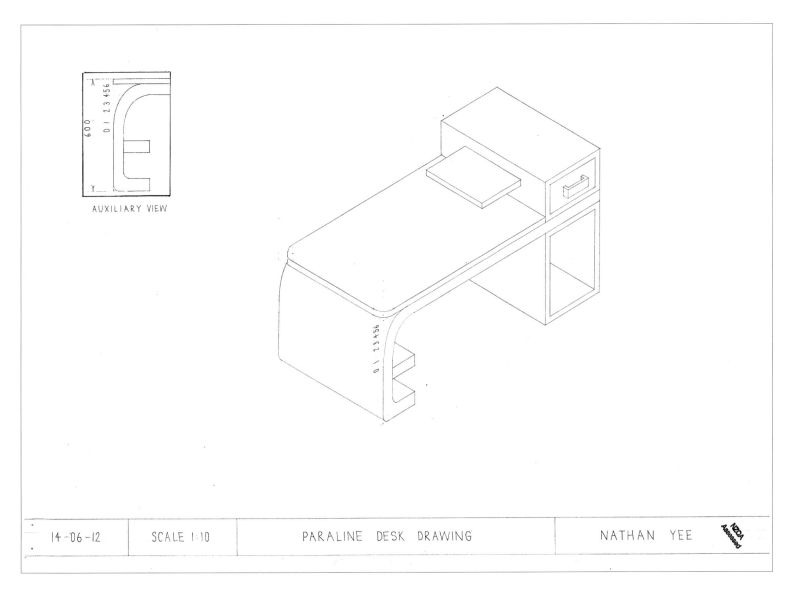

AUXILIARY VIEW

600

0 1 2 3 4 5 6

0 1 2 3 4 5 6

0 1 2 3 4 5 6

For this page, I used the same scale as the orthographic projection. It fitted perfectly when I turned it into an isometric drawing.

The time-consuming part was using ordinates to draw the curved end in 3D. By photocopying the curved end part of the front elevation though, I was able to draw it in 3D much quicker. But it was tricky plotting all the points with my compass.

I took time again to make sure I used the right line types, especially so all the constructions could be seen. I used my clutch pencil but had to keep the lead really sharp for good lines.

I'm glad I had done these types of drawings before. Otherwise I know I would have found this hard.

| 14-06-12 | SCALE 1:10 | PARALINE DESK DRAWING | NATHAN YEE | NZQA Assessed |

Belinda Van Eeden

Because I had curved parts on my design, I needed to use an auxiliary view to help draw them in 3D. I photocopied the part on the front elevation of my orthographic drawing that shows the curves, then pasted it onto the page.

I then drew an isometric crate to the right scale, and put ordinates onto the auxiliary view and isometric drawing to plot the curves.

I found this quite time consuming but it was the only way to get the curve from 2D to 3D. I was glad that I knew how to do this because of similar drawings I had done in class beforehand.

I used a French curve to outline all the curved parts to make sure my lines were smooth, using a 2H pencil.

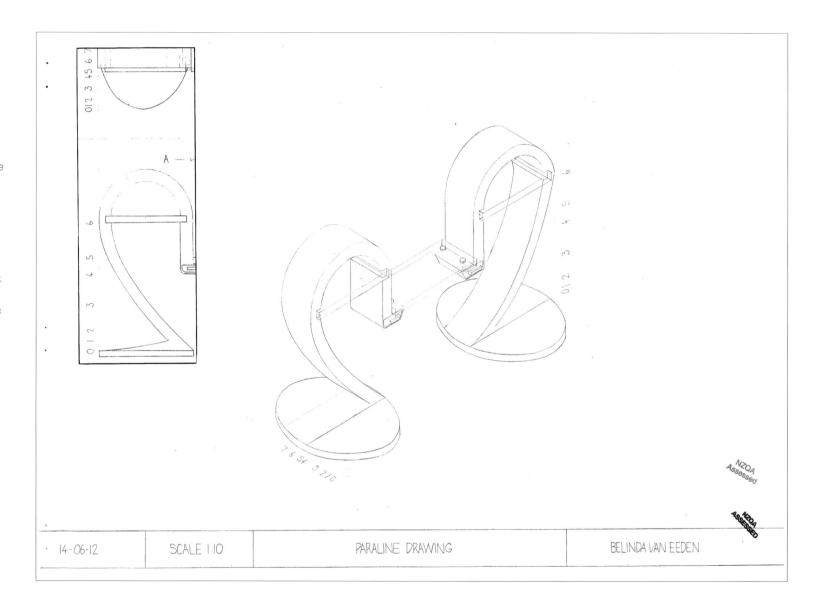

| 14-06-12 | SCALE 1:10 | PARALINE DRAWING | BELINDA VAN EEDEN |

ISBN: 9780170355575

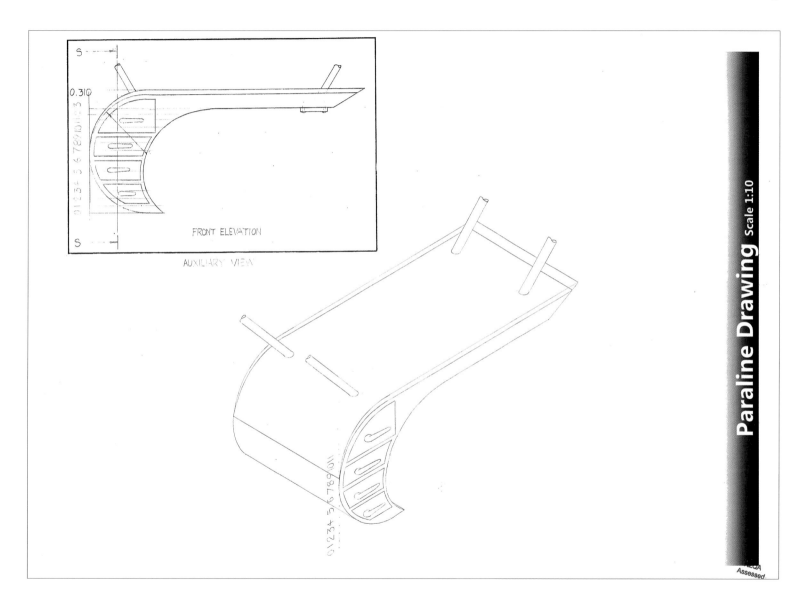

S

0.310

01234 5 6 789 10 1 2 3

FRONT ELEVATION

S

AUXILIARY VIEW

01234 5 6 789 10 1

Paraline Drawing Scale 1:10

Assessed

This drawing also took a lot of time, but once I had my crate drawn, it did not take as long as I thought it would.

First I photocopied the front elevation of my orthographic projection drawing so I could use it as an auxiliary view to plot the curves into 3D.

I put ordinates through this and numbered them, then drew a 3D crate the same size. I'm glad I knew about this method by doing drawings to practise on beforehand.

The hardest part was getting the angles of the steel rods right. But the auxiliary helped with this.

I wanted the lines to match my orthographic drawing, so I used a clutch pencil again with a 2H lead.

I used a French curve to outline the curved end and an ellipse template for the small circles at the end of the rods and drawer handles.

SITUATION

In your senior years at college you need a work space of your own at home.

BRIEF: Design a homework workstation.
DURATION: 20 weeks (class time)

REQUIREMENTS

1   Design research. *Cut and pasted among your sketches.*
2   Ideation of the workstation. Your sketches should show five stages: *thumbnails, exploratives, alternatives, thinking, and technical sketches. Consider human factors. Indicate your chosen design.* Four A3 pages minimum.
3   Place notes at any stage, about your research and sketches, to explain your thinking and details that are not clear visually.
4   Demonstrate in your sketches the effective development of your design ideas, *e.g. explore alternative ideas, not limited to only the brief and its requirements, then refine your chosen ideas informed by aesthetics and function. Support design judgements with qualitative and/or quantitative data gained through research, reflecting your values, tastes and/or views. Where appropriate, use drawings, models, digital modelling, etc.*
5   Draw with instruments, a detailed, scaled third angle orthographic projection of your workstation design. Show the following:
    ▪ A minimum of three views, fully dimensioned, with one elevation sectioned.
    ▪ Cutting planes, hidden detail, cross-hatching, notation and reference lines.
    ▪ A title block with the scale and projection symbol.
    ▪ Correct drawing standards and conventions.
6   Draw with instruments, to scale, the following:
    ▪ a paraline drawing, in either oblique or isometric, of your chosen design

    ▪ exploded isometric drawings of the main parts of the design to show construction detail.

7   Using media of your choice, produce a hand-rendered presentation drawing of your design that effectively communicates its shape and surface qualities.

DESIGN SPECIFICATIONS

The design must have curved or angled parts.
▪ Consideration of both aesthetics and function.
   Function: anthropometrics (chair and work surface height and sizes, etc.).
   Aesthetics: style, proportion, finish, harmony, form, etc.

Instrumental drawing (cont.)

EXPLODED PARALINE DRAWING

Here is the final instrumental drawing in the set.

To complement the previous drawings, an exploded drawing provides clear information that could allow the design to be manufactured. Again, it serves to showcase a range of student instrumental drawing ability, complementing the previous pages.

Four important factors need to be taken into account with this drawing:
▪ That the parts are exploded (pulled apart) in the direction in which they are assembled.
▪ That exploded lines and crates are clearly seen.
▪ That constructions of any curved or angled surfaces are clearly shown.
▪ If not labelled, the parts need to be identified in a parts list.

It is not necessary to draw an exploded view of the entire design of the kind found in kitset model making. Instead, a range of parts, important to the construction of the design, should be shown.

Knowledge of exploded drawing techniques, and how to make a parts list, must be taught. Note that the parts list can be produced on a computer, then cut and pasted onto the drawing.

As always, before these drawings can be attempted, comprehensive classwork exercises must be undertaken, directed at every step by the teacher. Paraline drawing (isometric and oblique) needs to be introduced in earlier years of study, so that students are not attempting to draw their designs without any prior knowledge.

Circles and curves that can be constructed with a compass are essential.

Note the addition of the auxiliary view (in this case, a photocopy of the curved part from the orthographic projection), which allows for the plotting of the compound curve part.

SKILLS NEEDED TO PRECEDE THIS PAGE:

Line weights and line standards (2H pencil for all lines).
Title block, scales and a parts list with numbered components.
Printing standards — 5 mm in title block, 3 mm about the drawing (HB pencil).
3D circle and curve constructions.
Links to: pages 71, 187–189

ISBN: 9780170355575

Nathan Yee

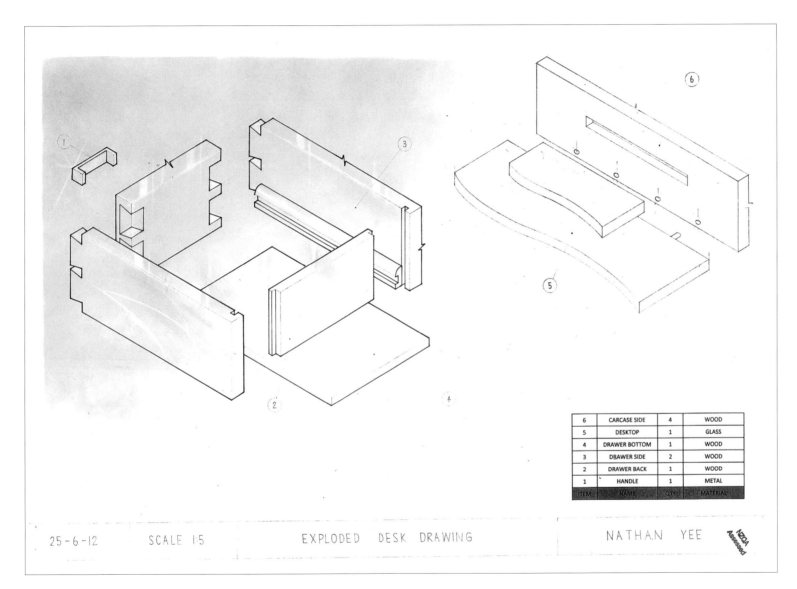

6	CARCASE SIDE	4	WOOD
5	DESKTOP	1	GLASS
4	DRAWER BOTTOM	1	WOOD
3	DRAWER SIDE	2	WOOD
2	DRAWER BACK	1	WOOD
1	HANDLE	1	METAL
ITEM	NAME	QTY	MATERIAL

25-6-12   SCALE 1:5   EXPLODED   DESK   DRAWING   NATHAN   YEE

I mainly had to work on getting the scales right for the exploded drawings and to make sure the proportions were right for fitting onto the page.

By having a parts list, it made it easier to see where I was going with the construction of the parts.

I knew I couldn't show all the parts of the desk, so I concentrated on the drawer and the shelf parts.

The research I had done earlier was helpful in finding out how the parts would be joined. The drawer was quite tricky as there are so many different types.

I used some thick and thin lines and a pastel background to make the page look more interesting.

Belinda Van Eeden

For this drawing I concentrated on showing some construction details of the drawer and a part of the desk.

Working out a scale to fit the page took some time, and also working out which parts to show exploded.

I did the parts list on a computer and used a circle template to draw the circles around the numbers on the drawing. This kept it neat. I used an ellipse template for the small circles and cylinder, and a French curve for the break symbols.

I was careful once again to show the best lines I could, and used a 2H pencil for these.

I did a practice exploded page first, to make sure I was on the right track.

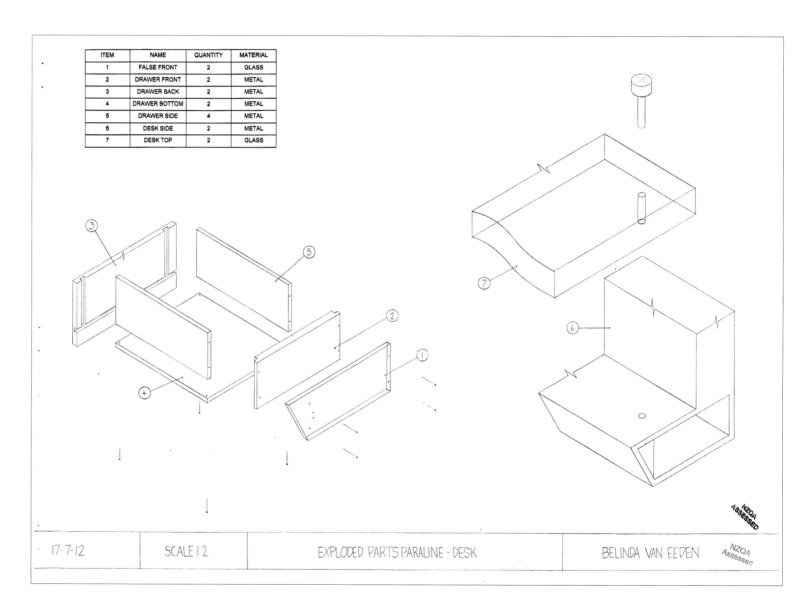

ITEM	NAME	QUANTITY	MATERIAL
1	FALSE FRONT	2	GLASS
2	DRAWER FRONT	2	METAL
3	DRAWER BACK	2	METAL
4	DRAWER BOTTOM	2	METAL
5	DRAWER SIDE	4	METAL
6	DESK SIDE	2	METAL
7	DESK TOP	2	GLASS

17-7-12	SCALE 1:2	EXPLODED PARTS PARALINE - DESK	BELINDA VAN EEDEN

ISBN: 9780170355575

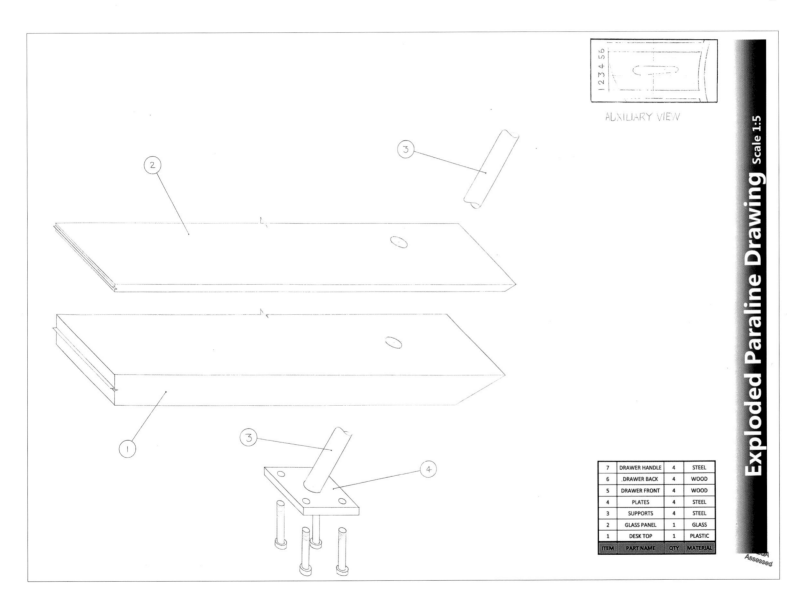

AUXILIARY VIEW

Exploded Paraline Drawing Scale 1:5

7	DRAWER HANDLE	4	STEEL
6	.DRAWER BACK	4	WOOD
5	DRAWER FRONT	4	WOOD
4	PLATES	4	STEEL
3	SUPPORTS	4	STEEL
2	GLASS PANEL	1	GLASS
1	DESK TOP	1	PLASTIC
ITEM	PART NAME	QTY	MATERIAL

Assessed

Out of all my instrumental drawings, this one took the longest time to do.

I decided to just show the steel rod parts and how they would attach to the desk.

The research I had done earlier was useful for this drawing. But the hardest part was getting the angles of them right in 3D.

I did this page three times before I got it right and the position on the paper. If I was to do it again I would move it across more to the right of the paper.

I used a clutch pencil and 2H lead again to keep the line work the same as my other drawings.

The small holes were drawn with an ellipse template.

I made the title on a computer, and the parts list, which was pasted onto the page.

In the end I was happy with the way the drawing turned out.

SITUATION

In your senior years at college you need a work space of your own at home.

BRIEF: Design a homework workstation.
DURATION: 20 weeks (class time)

REQUIREMENTS

1. Design research. *Cut and pasted among your sketches.*
2. Ideation of the workstation. Your sketches should show five stages: *thumbnails, exploratives, alternatives, thinking, and technical sketches. Consider human factors. Indicate your chosen design.* Four A3 pages minimum.
3. Place notes at any stage, about your research and sketches, to explain your thinking and details that are not clear visually.
4. Demonstrate in your sketches the effective development of your design ideas, *e.g. explore alternative ideas, not limited to only the brief and its requirements, then refine your chosen ideas informed by aesthetics and function. Support design judgements with qualitative and/or quantitative data gained through research, reflecting your values, tastes and/or views. Where appropriate, use drawings, models, digital modelling, etc.*
5. Draw with instruments, a detailed, scaled third angle orthographic projection of your workstation design. Show the following:
   - A minimum of three views, fully dimensioned, with one elevation sectioned.
   - Cutting planes, hidden detail, cross-hatching, notation and reference lines.
   - A title block with the scale and projection symbol.
   - Correct drawing standards and conventions.
6. Draw with instruments, to scale, the following:
   - a paraline drawing, in either oblique or isometric, of your chosen design
   - exploded isometric drawings of the main parts of the design to show construction detail.

7. Using media of your choice, produce a hand-rendered presentation drawing of your design that effectively communicates its shape and surface qualities.

DESIGN SPECIFICATIONS

The design must have curved or angled parts.
- Consideration of both aesthetics and function.
  Function: anthropometrics (chair and work surface height and sizes, etc.).
  Aesthetics: style, proportion, finish, harmony, form, etc.

Presentation drawing

Here is the final drawing in the set. And it's time to bring out the big guns! This is the drawing that will 'sell' the design, so it must have the 'wow' factor. Careful consideration of viewpoint is essential to achieving this.

Firstly look back at the design sketches. There may be one that already has the right look and feel. If so, re-sketch it freehand, neater, and with cleaner linework. *By sketching the presentation drawing freehand, more freedom of line and a more 'styley' look can be achieved, rather than a 'clunky' look that comes from an instrumental drawing or an isometric or oblique drawing. Perspective works best with exaggerated lines.*

Make several photocopies on cartridge paper (not thin photocopy paper) to allow for the use of marker pens. Photocopying will also enable the drawing to be enlarged or reduced to suit. Preserve one copy as the master, on which final renderings and cut and pasted areas can be placed. Use the others for practice. Once the master is fully rendered and outlined, it can be cut out ready to be mounted onto a background.

Nathan and Stephan have worked mainly with marker pen and chalk pastel, while Belinda has chosen the airbrush to better define the curved surfaces. White fineline gel pens add edge highlights and gouache adds sparkle to the corners. Background and shadow have been produced by smudging chalk pastel pencil within a masked rectangle drawn on a toothed card, with edges sharpened with an eraser. Eraser stripes also add shine to the glass surfaces. White gel pen has been used to define edges of the backgrounds and shadow, and sparkle to the glass edges caught by the light.

Remember that these techniques must be taught — they are not a given. Precede this drawing, as always, with exercises that give practice in applying a range of media to achieve different surface textures and form that will ultimately inform decisions about students' own designs.

Also practise drawing backgrounds that complement the style of the drawing while at the same time grounding the sketch on the page. Careful consideration of colour for the background, so that the eye goes to the sketch, is important.

SKILLS NEEDED TO PRECEDE THIS PAGE:

Perspective drawing techniques.
Viewpoint, backgrounds.
Crating (proportions) and line hierarchy.
Rendering techniques — light source, tonal changes, shape and surface texture.
Colour palette considered.
Media technique: marker pens, black fineline pens, pastel pencils.
Links to: pages 82–91

Hint: Much individual advice from the teacher is required at this time, so sound class management is needed. And, finally, that 'styley' signature practised earlier can now be placed on the drawing.

ISBN: 9780170355575

I liked doing this page although it took a long time. I wanted it to be innovative and creative, so I explored using different rendering methods than my design pages.

Instead of using my isometric drawing to render, I redrew my desk freehand, in two-point perspective, and on a steeper angle to get a better look and to show off the rounded end and glass top.

I rendered it with marker pens to show the wood. I then cut it out and pasted it onto the background, which I did with chalk pastels smudged inside a mask of Frisket Film.

The glass top was outlined in white gel pen, then smudged with white pastel pencil over the marker pen and background. I used white gouache to give the desk sparkle.

Belinda Van Eeden

This drawing took a lot of time. I had to think carefully about how best to show the materials it was made from, and how to make the design look good on the page.

I photocopied my paraline drawing and used this to render. I used a new rendering method, which was the airbrush, to show the curved ends. I had one of these in my classroom, so I could practise first. The bottom parts were rendered using a grey marker pen with smudged white pastel pencil. I used my eraser to make the stripes on the top for a shiny look.

When the end parts were rendered, I cut them out and pasted them onto a dark card background using spray glue. I then drew the glass shelf onto the background with a white gel pen, and smudged white pastel on top to give the shiny see-through effect ... I was really pleased with the way it looked.

The shadow underneath is combinations of different colours of smudged chalk pastel. I used a white gel pen to make the highlights and add sparkle to the edge of the glass.

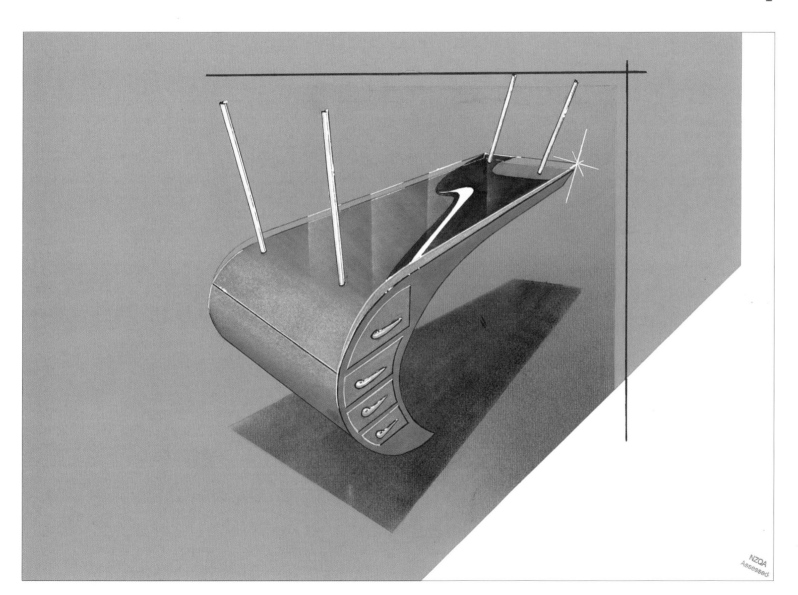

For this drawing I wanted it to look stylish, so I used the same viewpoint as the one on my explorative sketch page.

The hardest part was deciding on what rendering media to use. I was going to airbrush it but used marker pen, with white pastel pencil smudged onto the rounded end.

I researched how I could show the glass top and found that black marker pen with white looks good.

When I had rendered the main part of the desk, I cut it out and pasted it onto a card background. Then I used a 0.01 black fineline pen to draw the outline of the glass top onto the background and the top of the desk.

The vertical stripes of shine on the top were done with a white pastel pencil smudged against an erasing shield.

The background and shadow was done by smudging pastel pencil into the card background.

I added sparkle to the top back corner with gouache and a white pen.

SPATIAL DESIGN

SITUATION

Traditionally and culturally, Kiwi summer holidays take place in the great outdoors. It's a time to experience the laid-back nature of New Zealand life and travel to hidden corners of our country.

BRIEF: Design a mobile home and a feature for the home.
DURATION: 20 weeks (class time)

REQUIREMENTS   **PART ONE: SPATIAL DESIGN**

1. Design research. *Cut and pasted among your sketches.*
2. Ideation of the exterior of the home. *Show thumbnails, explorative sketches and alternative ideas. Three A3 pages minimum.*

3. Two bubble diagrams to show planning of the interior.

4. Place notes at any stage, about your research and sketches, to explain thinking and details that are not clear visually. *Discuss: sustainable building materials, energy generation (solar power/heating, etc.), waste management.*
5. Demonstrate in your sketches the effective development of your design ideas, *e.g. explore alternative ideas, not limited to only the brief and its requirements, then refine your chosen ideas informed by aesthetics and function. Support design judgements with qualitative and/or quantitative data gained through research, reflecting your values, tastes and/or views. Where appropriate, use drawings, models, digital modelling, etc.*

6. Draw with instruments, a detailed, scaled third angle orthographic projection of your mobile home design. Show the following:
   - A minimum of three views, fully dimensioned. *Draw the plan as a floor plan to show interior layout, with one elevation either a part section or full section, with cutting plane and cross-hatching.*
   - Hidden detail, cross-hatching, notation and reference lines.
   - A title block with the scale and projection symbol.
   - Correct drawing standards and conventions.
7. Using media of your choice, produce a hand-rendered presentation drawing of your design, in a landscape setting, that effectively communicates its shape and surface qualities.

DESIGN SPECIFICATIONS

The home should:
- be designed with sustainability in mind and accommodate the user in comfort
- be able to be moved when needed — either on wheels or lifted to a new site
- consider both aesthetics and function.
  Function: construction, safety, user friendliness, fit for purpose, ergonomics, etc.
  Aesthetics: style, proportion, finish, harmony, form, etc.

Ideation

THUMBNAILS AND EXPLORATIVE SKETCHES

The intention of this brief is to cover both areas (spatial and product design) from one unit of work: a mobile home design (spatial design), and then a feature that could be used in, or be a part of, the mobile home (product design).

    The unit provides the student with a wide range of learning experiences and important assessment opportunities.

    A sophisticated sketching style must be encouraged at this level.

    Careful attention, on the featured drawings, has been given to layout, viewpoints, colour palette and page background.

    Rendering is intentionally minimalist, working with colours that provide information about form and materials. A shadow beneath sketches provides depth, while good use of black ballpoint pen on craft paper enhances the thumbnails.

    Ensuring a mix of 2D and 3D sketches and different viewpoints is an important consideration.

    Titles have been produced on a computer, printed onto craft paper and pasted onto the pages or done by hand.

SKILLS NEEDED TO PRECEDE THIS PAGE:

A planning sheet.

Thumbnailing techniques on craft paper/media use.

Freehand design sketching in 2D and 3D with clearly seen crating (proportions) and line hierarchy.

Viewpoints are considered for visual interest and to better show the ideas.

Arrows indicate movement and/or function.

Rendering techniques — minimalist, indicating the light source, tonal changes and form.

Colour palette — soft grey tones, light blues and black.

Links to: pages 75–81

ISBN: 9780170355575

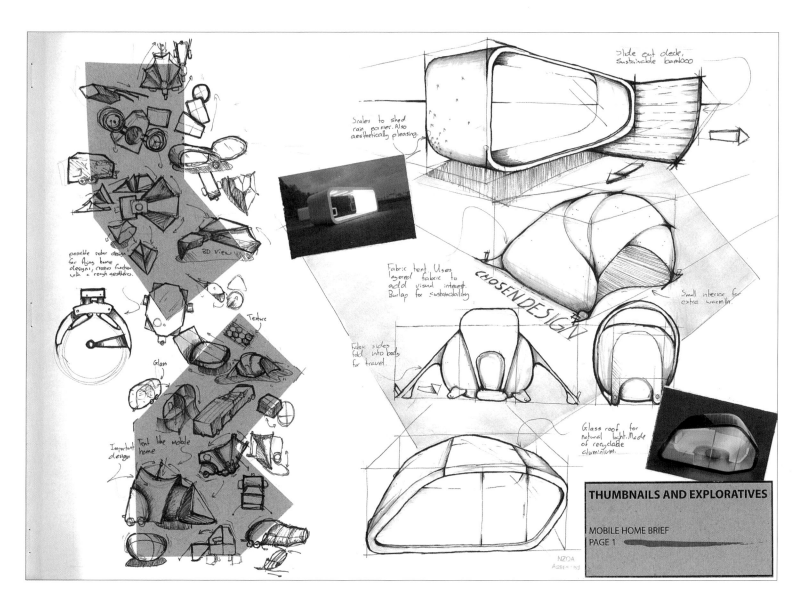

Slide out deck, sustainable bamboo

Scales to shed rain easier. Also aesthetically pleasing.

Fabric tent. Uses layered fabric to add visual interest. Burlap for sustainability

Fabric sides fold into body for travel.

CHOSEN DESIGN

Small interior for extra warmth.

Glass roof for natural light. Made of recyclable aluminium.

possible solar design for flying home designs, crosses fashion with a rough aesthetic.

3D View

Texture

Glass

Important design

Tent like mobile home

THUMBNAILS AND EXPLORATIVES

MOBILE HOME BRIEF
PAGE 1

NZQA Assessed

I wanted to try more 'sketchy' but stylish type of sketches for his brief, but the challenge was to make sure I could show the details of my designs ... I spent a lot of time practising sketches first to get the look I wanted.

I planned the layout first but had to concentrate on setting out the crates to make sure I had the best viewpoints to show the parts of the designs well. This was the hardest part and took a while.

I did the thumbnails first, using a black ballpoint pen. I like how you can show shading and dark and light lines with a ballpoint.

I kept my explorative sketches simple but tried different viewpoints in 2D and 3D. They were outlined with thick and thin lines using a black fineline pen. I deliberately used soft pastel colours to render them using a marker pen.

If I was to do the page again, I would make some of the lines bolder.

Nicole Dealey

This was the first page I did for the year so I was careful to get a good layout and the right kinds of sketches. I did a practice layout first.

I tried a lot of different viewpoints for my mobile home ideas and decided to show five on the page. Four I did in 3D and one in 2D. Getting the porportions right for the crates took the longest because I didn't know how big I was going to make the design at first.

I used shadows under some of the views to make them stand out more on the page.

I kept the rendering greyscale, using markers and black fineline pens to outline, but would have used a bit more colour if I was to do the page again.

I would also make the thick line outlines thicker in future to make them stand out more from the other lines.

I did the title block on a computer and printed them onto the page before I started drawing.

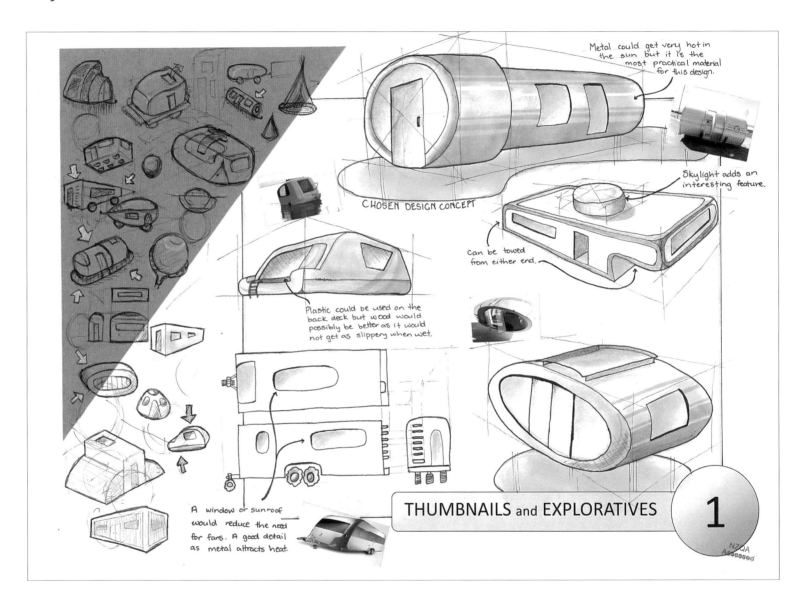

Metal could get very hot in the sun but it is the most practical material for this design.

Skylight adds an interesting feature.

CHOSEN DESIGN CONCEPT

Can be towed from either end.

Plastic could be used on the back deck but wood would possibly be better as it would not get as slippery when wet.

A window or sunroof would reduce the need for fans. A good detail as metal attracts heat.

THUMBNAILS and EXPLORATIVES   1

NZQA Assessed

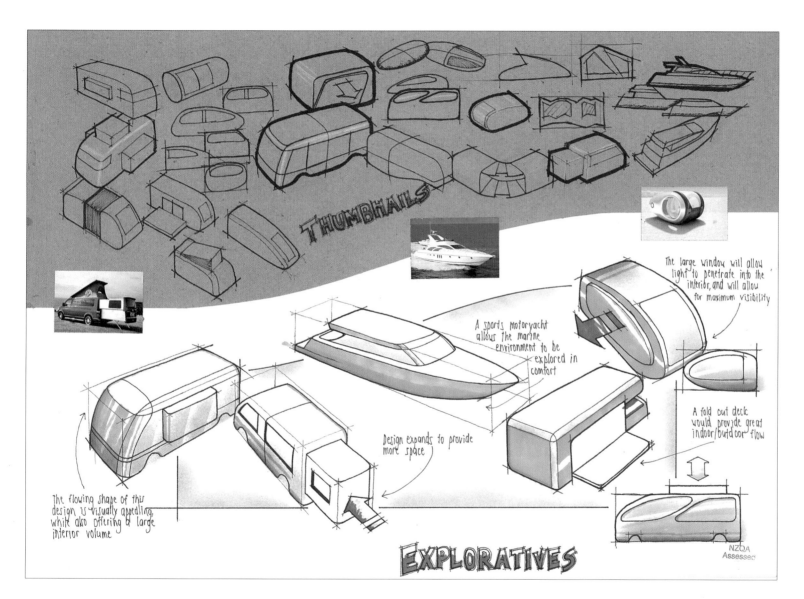

I thought a lot about this page before I did it, as I wanted to show a lot of information and detail in my drawings.

I planned all parts of the page beforehand to get the look I wanted. I had done thumbnails down the side before, but decided to put them across the top to leave better space for my designs, which were long instead of tall.

The hardest part was deciding on different viewpoints for the sketches and the proportions for the crates. Although all of them are looking down on the top, if I was to do them again, I would do some looking more into the side. But I like the way they turned out.

I kept the rendering simple, using grey marker pen on the dark surfaces of the designs. The background is light-blue smudged pastel pencil to blend in with the grey colour of the sketches.

I pasted research on the curve of my thumbnail paper to show the flow of my thinking.

SITUATION

Traditionally and culturally, Kiwi summer holidays take place in the great outdoors. It's a time to experience the laid-back nature of New Zealand life and travel to hidden corners of our country.

BRIEF: Design a mobile home and a feature for the home.
DURATION: 20 weeks (class time)

REQUIREMENTS	PART ONE: SPATIAL DESIGN

1   Design research. *Cut and pasted among your sketches.*
2   Ideation of the exterior of the home. *Show thumbnails, explorative sketches and alternative ideas. Three A3 pages minimum.*

3   Two bubble diagrams to show planning of the interior.

4   Place notes at any stage, about your research and sketches, to explain thinking and details that are not clear visually. *Discuss: sustainable building materials, energy generation (solar power/heating, etc.), waste management.*
5   Demonstrate in your sketches the effective development of your design ideas, *e.g. explore alternative ideas, not limited to only the brief and its requirements, then refine your chosen ideas informed by aesthetics and function. Support design judgements with qualitative and/or quantitative data gained through research, reflecting your values, tastes and/or views. Where appropriate, use drawings, models, digital modelling, etc.*

6   Draw with instruments, a detailed, scaled third angle orthographic projection of your mobile home design. Show the following:
   ▪ A minimum of three views, fully dimensioned. *Draw the plan as a floor plan to show interior layout, with one elevation either a part section or full section, with cutting plane and cross-hatching.*
   ▪ Hidden detail, cross-hatching, notation and reference lines.
   ▪ A title block with the scale and projection symbol.
   ▪ Correct drawing standards and conventions.
7   Using media of your choice, produce a hand-rendered presentation drawing of your design, in a landscape setting, that effectively communicates its shape and surface qualities.

DESIGN SPECIFICATIONS

The home should:
▪ be designed with sustainability in mind and accommodate the user in comfort
▪ be able to be moved when needed — either on wheels or lifted to a new site
▪ consider both aesthetics and function.
   Function: construction, safety, user friendliness, fit for purpose, ergonomics, etc.
   Aesthetics: style, proportion, finish, harmony, form, etc.

Ideation (cont.)

ALTERNATIVES AND THINKING SKETCHES

At this stage, after initial exploration of ideas, students need to be given the opportunity to take their thinking further and explore alternatives that could better inform the design.

They need to be encouraged to look elsewhere for inspiration and to think 'outside the square'.

A lesson on bio-mimickry provokes thinking in the right direction. Anton's research of shapes in nature are clearly linked to an emerging train of thought that is better informing his thinking.

Again, a simple yet sophisticated sketching style can be seen on this page, with rendering only applied to highlight tone, surface shape and shadows.

His work with grey-tone markers and black fineline pens conveys all that is required to know at this stage. Lots of white space allows his sketching to 'breathe' on the page.

 SKILLS NEEDED TO PRECEDE THIS PAGE:

A planning sheet.
Freehand design sketching in 2D and 3D with clearly seen crating (proportions) and line hierarchy.
Viewpoints are considered for visual interest and to better show the ideas.
Arrows indicate movement and/or function.
Rendering techniques — minimalist, indicating the light source, tonal changes and form.
Colour palette — soft grey tones, light blues and black.
Links to: pages 75–81

ISBN: 9780170355575   PHOTOCOPYING OF THIS PAGE IS RESTRICTED UNDER LAW.

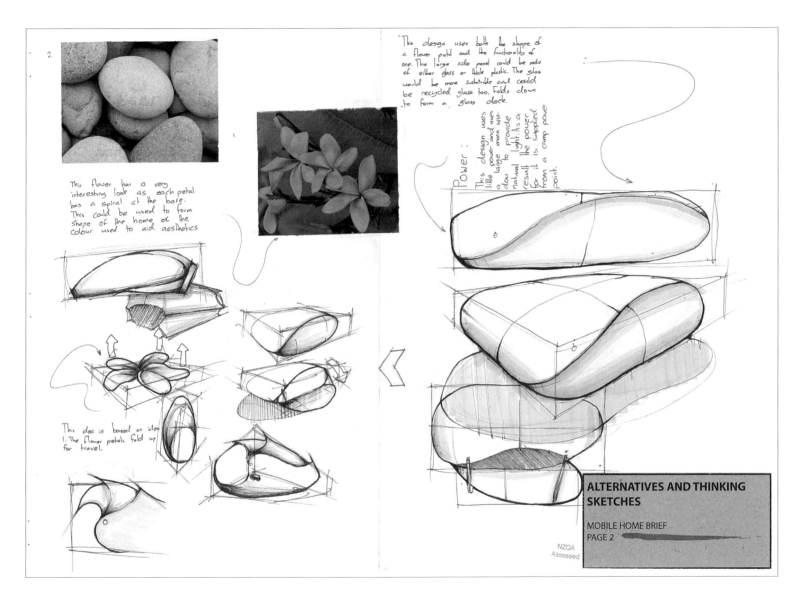

2.

This flower has a very interesting look as each petal has a spiral at the base. This could be used to form shape of the home or the colour used to aid aesthetics

This idea is based on idea 1. The flower petals fold up for travel.

This design uses both the shape of a flower petal and the functionality of one. The large side panel could be made of either glass or thick plastic. The glass would be more sustainable and could be recycled glass too. Folds down to form a glass deck.

Power: This design uses little power and uses a large main window to provide natural light. As a result the power for it is supplied from a camp power point.

ALTERNATIVES AND THINKING SKETCHES

MOBILE HOME BRIEF
PAGE 2

NZQA
Assessed

I wished I had spent more time planning this page first. I found some good alternative research but I think my sketches lack coherency on the left side of the page.

The sketches on the right side are better I think, because they are simple, and you can see the influence of the rocks and leaves in the design ... I started to think about function more and put some arrows to show how the drop-down side of my idea might work.

I looked back at my first page and tried to keep the style the same. I used thick and thin lines again to outline the sketches and drew some shadows to make them more effective. The same rendering as on my first page was used.

At this stage I sorted out my title blocks and printed them on a computer and glued them onto the page..

Nicole Dealey

For this page I wanted to look at different objects to give me a better idea for the design. I chose a watch, camera lens and a trombone because I liked the rounded parts of each.

I chose to draw mainly in perspective as this gives the drawings a more realistic look. I got the idea for the viewpoint and the design for the bottom drawing from the camera lens research. I like the way this drawing looks and think it is one of my better sketches.

I planned the page layout first and used brown paper again to match my first page.

For the rendering I used grey marker pen. This was easy because my designs are mainly cylinders and I knew how to render them from exercises I had done in class.

If I was to do the page again I would make my chosen design bigger and with more bolder rendering.

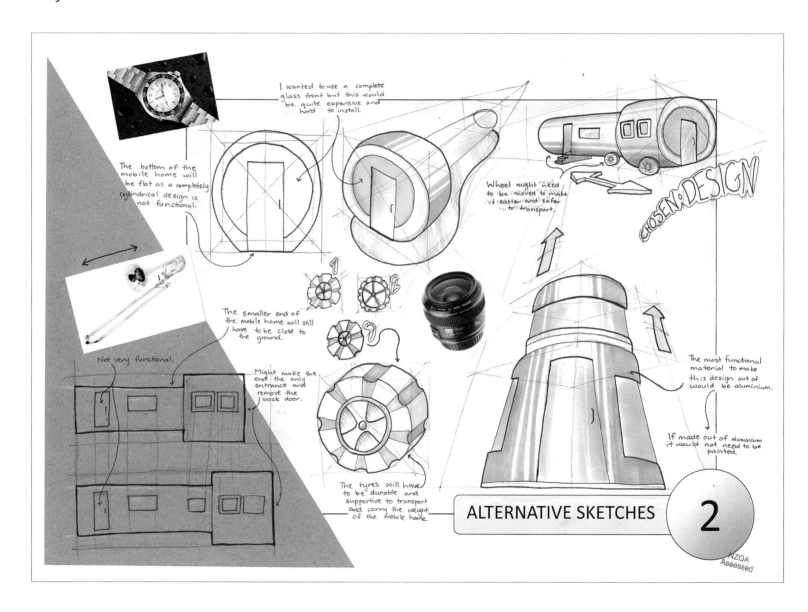

I wanted to use a complete glass front but this would be quite expensive and hard to install.

The bottom of the mobile home will be flat as a completely cylindrical design is not functional.

Wheel might need to be moved to make it easier and safer to transport.

CHOSEN DESIGN

The smaller end of the mobile home will still have to be close to the ground.

Not very functional.

Might make the end the only entrance and remove the back door.

The tyres will have to be durable and supportive to transport and carry the weight of the mobile home.

The most functional material to make this design out of would be aluminium.

If made out of aluminium it would not need to be painted.

ALTERNATIVE SKETCHES 2

NZQA Assessed

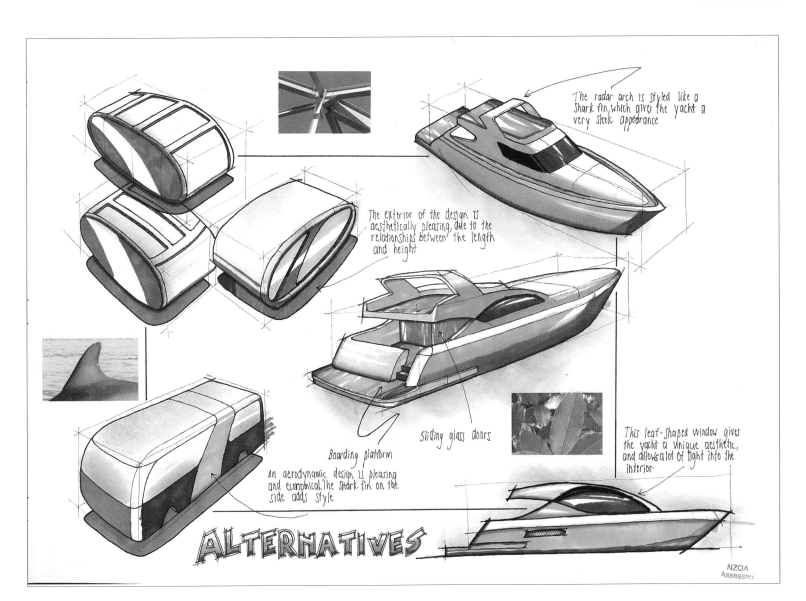

The radar arch is styled like a shark fin, which gives the yacht a very sleek appearance

The exterior of the design is aesthetically pleasing, due to the relationships between the length and height

Sliding glass doors

Boarding platform

An aerodynamic design is pleasing and economical. The shark fin on the side adds style

This leaf-shaped window gives the yacht a unique aesthetic, and allows a lot of light into the interior

ALTERNATIVES

NZQA Assessor

I'm glad that I had prior sketching skills that helped me complete this page, as my designs were becoming quite complex at this stage.

I was also glad to have had a good understanding of perspective drawing from my Year 10 work, as these became my favourite way of sketching my ideas.

I really like the way this page turned out, as the drawings look quite real. I got a good idea by using the fin of a shark to give the design a sleeker look.

I practised a lot of rendering techniques before I did this page. This helped in making sure the materials the designs are made from look realistic. I used marker pens again, and the same smudged pastel technique for the background.

I kept the colour palette of the page the same as my first page, and used the same style of lettering for the title.

SITUATION

Traditionally and culturally, Kiwi summer holidays take place in the great outdoors. It's a time to experience the laid-back nature of New Zealand life and travel to hidden corners of our country.

BRIEF: Design a mobile home and a feature for the home.
DURATION: 20 weeks (class time)

REQUIREMENTS   PART ONE: SPATIAL DESIGN

1 Design research. *Cut and pasted among your sketches.*
2 Ideation of the exterior of the home. *Show thumbnails, explorative sketches and alternative ideas. Three A3 pages minimum.*

3 Two bubble diagrams to show planning of the interior.

4 Place notes at any stage, about your research and sketches, to explain thinking and details that are not clear visually. *Discuss: sustainable building materials, energy generation (solar power/heating, etc.), waste management.*
5 Demonstrate in your sketches the effective development of your design ideas, *e.g. explore alternative ideas, not limited to only the brief and its requirements, then refine your chosen ideas informed by aesthetics and function. Support design judgements with qualitative and/or quantitative data gained through research, reflecting your values, tastes and/or views. Where appropriate, use drawings, models, digital modelling, etc.*

6 Draw with instruments, a detailed, scaled third angle orthographic projection of your mobile home design. Show the following:
- A minimum of three views, fully dimensioned. *Draw the plan as a floor plan to show interior layout, with one elevation either a part section or full section, with cutting plane and cross-hatching.*
- Hidden detail, cross-hatching, notation and reference lines.
- A title block with the scale and projection symbol.
- Correct drawing standards and conventions.
7 Using media of your choice, produce a hand-rendered presentation drawing of your design, in a landscape setting, that effectively communicates its shape and surface qualities.

DESIGN SPECIFICATIONS

The home should:
- be designed with sustainability in mind and accommodate the user in comfort
- be able to be moved when needed — either on wheels or lifted to a new site
- consider both aesthetics and function.
  Function: construction, safety, user friendliness, fit for purpose, ergonomics, etc.
  Aesthetics: style, proportion, finish, harmony, form, etc.

Ideation (cont.)

TECHNICAL SKETCHES

Students are not expected to know construction details of their mobile home, so ideation only needs to cover thumbnails and explorative sketches of the exterior.

However, scope is provided for students to explore this detail if they wish, as Anton has done. The full range of design sketches will be shown in product design.

Technical sketches need to be a mix of 3D exploded isometric and 2D sectioned views of parts. Anton has showed this well in his ideas for a pull-down side of his home.

Human factors are also introduced when looking at a hand grip.

Note the good use of line hierarchy (thick and thin lines) that provide the sketches with visual impact, and the use of crating to allow accurate construction of the circular and curved parts.

Research is ongoing, continuously informing ideas. Again, rendering is only applied to highlight tone, surface shape and shadows. The use of craft paper and a pastel background provides visual interest to the page. Lots of white space allows his sketching to 'breathe' on the page.

At this stage he is able to indicate his chosen design.

SKILLS NEEDED TO PRECEDE THIS PAGE:

Planning sheet used.

Freehand design sketching in 2D and 3D with clearly seen crating (proportions) and line hierarchy.

Viewpoints are considered for visual interest and to better show the ideas.

Rendering techniques — minimalist, indicating the light source, tonal changes and form.

Colour palette — soft grey tones, light blues and black.

Links to: pages 75–81

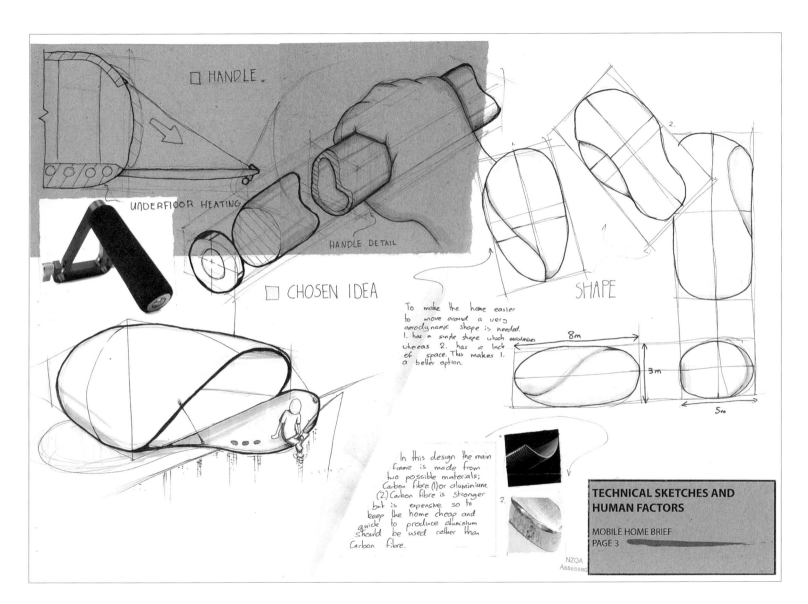

I found this page quite difficult at first, because I had to think about how parts could work.

But once I did some research, it was then a matter of working out what types of sketches would show things best.

The first exploded drawing of the handle I did was too small and didn't have the right look, so I started again. The brown paper is pasted over the top to cover it up. I was pleased with the way this turned out and how clear the new sketch is. I found that larger sketches are better than small ones to show inside detail.

I used different viewpoints again on the other sketches to make the page more interesting to look at. Working out the sizes helped me get things more in proportion.

I used marker pens and pastel colours again for the rendering with thick and thin outlines done with fineline black pens to match my other pages.

Nicole Dealey

Once I had my chosen design, I discovered that I needed to find out how it could be made for doing some technical sketches. But I found this to be hard, so I just focused instead on some of the materials and human factors to get sizes.

It took a while to work out the sizes. I measured a caravan and used similar sizes and also got the way my design could be towed.

I kept all the sketches simple, mainly 2D, and rendered them with grey marker pen again.

I researched materials on the internet and did a write-up of them using brown paper again to match the other pages. These were the only notes that I wrote in detail. The other ones I kept simple because I feel all the information is explained in my drawings.

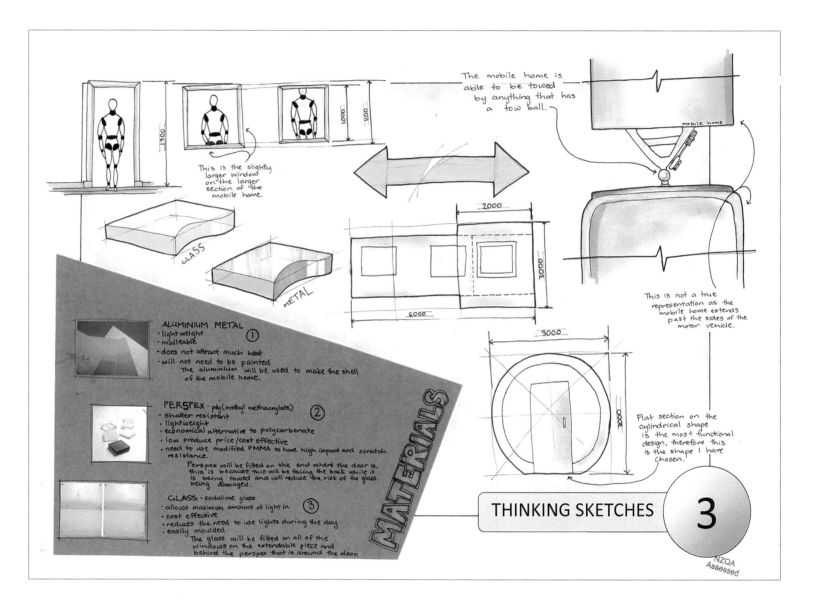

This is the slightly larger window on the larger section of the mobile home.

The mobile home is able to be towed by anything that has a tow ball.

This is not a true representation as the mobile home extends past the sides of the motor vehicle.

Flat section on the cylindrical shape is the most functional design, therefore this is the shape I have chosen.

ALUMINIUM METAL ①
· light weight
· malleable
· does not attract much heat
· will not need to be painted
The aluminium will be used to make the shell of the mobile home.

PERSPEX - poly(methyl methacrylate) ②
· shatter resistant
· lightweight
· economical alternative to polycarbonate
· low produce price/cost effective
· need to use modified PMMA to have high impact and scratch resistance.
Perspex will be fitted on the end where the door is, this is because this will be facing the back while it is being towed and will reduce the risk of the glass being damaged.

GLASS - sodalime glass ③
· allows maximum amount of light in
· cost effective
· reduces the need to use lights during the day
· easily moulded
The glass will be fitted on all of the windows on the extendable piece and behind the perspex that is around the door.

MATERIALS

THINKING SKETCHES 3

NZQA Assessed

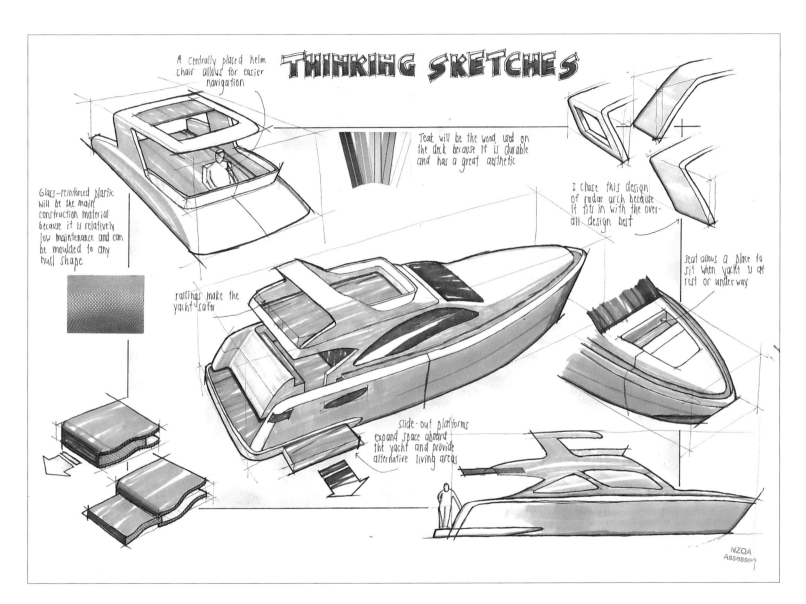

THINKING SKETCHES

A centrally placed helm chair allows for easier navigation

Teak will be the wood used on the deck because it is durable and has a great aesthetic

I chose this design of radar arch because it fits in with the overall design best

Glass-reinforced plastic will be the main construction material because it is relatively low maintenance and can be moulded to any hull shape

railings make the yacht safer

seat allows a place to sit when yacht is at rest or underway

slide-out platforms expand space aboard the yacht and provide alternative living areas

NZQA Assessor

I spent a lot of time planning out the layout for this page on scrap paper.

Firstly, I planned where the dominant image would be placed, then what angle it would be viewed from.

I wanted the design to be clearly seen, with specific parts and details of the design shown in smaller sketches around the dominant image.

I made this page a thinking sketches page as I was not too sure about how to show the construction in technical sketches. I realised at this stage that I had chosen a fairly complex design to draw. This made the page a hard one to do.

The part I liked the most was the rendering, which I did with marker pens. I really like how it turned out, especially the wood parts, which I did with a colour called sandstone.

I did a smudged pastel background to match my other pages and to guide the eye around my sketches.

SITUATION

Traditionally and culturally, Kiwi summer holidays take place in the great outdoors. It's a time to experience the laid-back nature of New Zealand life and travel to hidden corners of our country.

BRIEF: Design a mobile home and a feature for the home.
DURATION: 20 weeks (class time)

REQUIREMENTS — PART ONE: SPATIAL DESIGN

1. Design research. *Cut and pasted among your sketches.*
2. Ideation of the exterior of the home. *Show thumbnails, explorative sketches and alternative ideas. Three A3 pages minimum.*

3. Two bubble diagrams to show planning of the interior.

4. Place notes at any stage, about your research and sketches, to explain thinking and details that are not clear visually. *Discuss: sustainable building materials, energy generation (solar power/heating, etc.), waste management.*
5. Demonstrate in your sketches the effective development of your design ideas, *e.g. explore alternative ideas, not limited to only the brief and its requirements, then refine your chosen ideas informed by aesthetics and function. Support design judgements with qualitative and/or quantitative data gained through research, reflecting your values, tastes and/or views. Where appropriate, use drawings, models, digital modelling, etc.*
6. Draw with instruments, a detailed, scaled third angle orthographic projection of your mobile home design. Show the following:
   - A minimum of three views, fully dimensioned. *Draw the plan as a floor plan to show interior layout, with one elevation either a part section or full section, with cutting plane and cross-hatching.*
   - Hidden detail, cross-hatching, notation and reference lines.
   - A title block with the scale and projection symbol.
   - Correct drawing standards and conventions.
7. Using media of your choice, produce a hand-rendered presentation drawing of your design, in a landscape setting, that effectively communicates its shape and surface qualities.

DESIGN SPECIFICATIONS

The home should:
- be designed with sustainability in mind and accommodate the user in comfort
- be able to be moved when needed — either on wheels or lifted to a new site
- consider both aesthetics and function.
  Function: construction, safety, user friendliness, fit for purpose, ergonomics, etc.
  Aesthetics: style, proportion, finish, harmony, form, etc.

Planning of spaces

BUBBLE DIAGRAMS

This page shows the spatial relationships of the inside of the chosen mobile home design. Two layouts, as asked for in the brief, are drawn using bubble diagrams.

The chosen layout is then converted into a line diagram (the footprint shape of the chosen home design). Note how the spaces inside the home have been listed so that none is missed out when converting to the bubble diagrams.

Anton has made good use of landscape graphics techniques, with an eye to presentation of the page.

Drawing of the bubbles, their linking arrows, and landscape graphics in plan view, need to be taught to students. The bubbles have been drawn in pencil with a circle template first, then inked in with fineline pens.

Again, Anton's sophisticated sketching style enhances this page, with minimalist rendering being applied.

Grey-tone and light-blue markers preserve the colour palette of his ideation pages, while use of a sandstone marker on the landscape features provides a focus.

A subtle, smudged blue pastel background anchors the bubbles at the top left of the page.

SKILLS NEEDED TO PRECEDE THIS PAGE:

A planning sheet.
Bubble drawing techniques and linking arrows. Architectural features in plan view.
Rendering techniques — minimalist, indicating the light source, tonal changes and form.
Colour palette — soft grey tones, light blues and black.
Links to: pages 27–29, 76

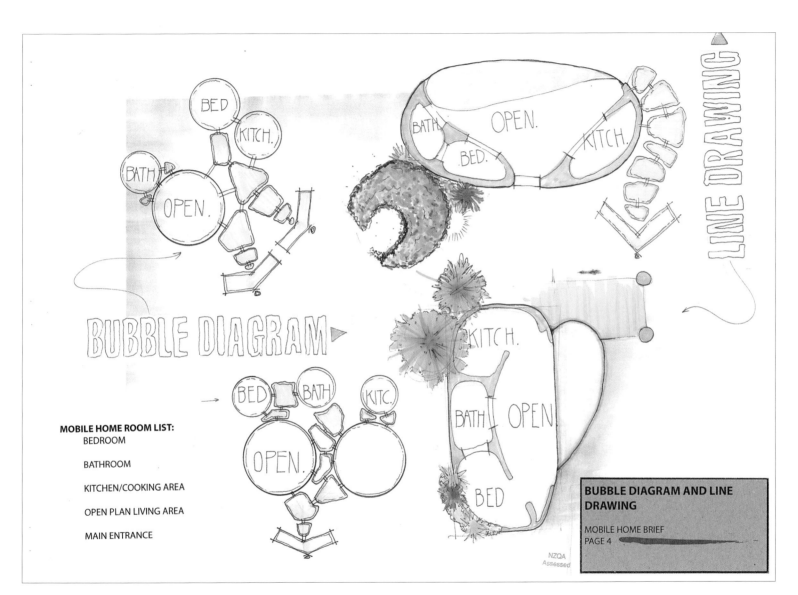

MOBILE HOME ROOM LIST:

BEDROOM

BATHROOM

KITCHEN/COOKING AREA

OPEN PLAN LIVING AREA

MAIN ENTRANCE

BUBBLE DIAGRAM

LINE DRAWING

BUBBLE DIAGRAM AND LINE DRAWING

MOBILE HOME BRIEF
PAGE 4

NZQA
Assessed

I deliberately moved away from the style of my previous pages for this page.

Here I am looking at the planning of the interior of the mobile home, based on my chosen design.

I used a landscape bubble diagram style, rather than a more classic interior layout style, as I wanted to keep the page looking styley like my others. I also prefer this look.

I set out the bubbles using a circle template, then inked them in with a black fineline pen.

For the style of the arrows and plants, I researched first in landscape architecture books and online, then used the ideas I liked the best on my drawing.

I took care also with the printing to keep it an 'architectural' look.

Although some parts of the page did not come out as expected, the look still works and I am pleased with the finished page.

Nicole Dealey

This was a fun page where I looked at how the inside of my design could be arranged.

I used a circle and an ellipse template for the bubbles and rendered them with marker pen. The outlines were done with a 0.1 and a 0.3 black fineline pen.

I researched architectural styles for arrows and lettering beforehand, and this helped give the page a more stylish look.

I decided to use brown paper again to match the other pages, and drew the line diagrams on this to contrast with the bubble diagrams.

If I was to do the page again, I would have printed out a computer title to match the other pages instead of doing it by hand. But I like the way it turned out in the end.

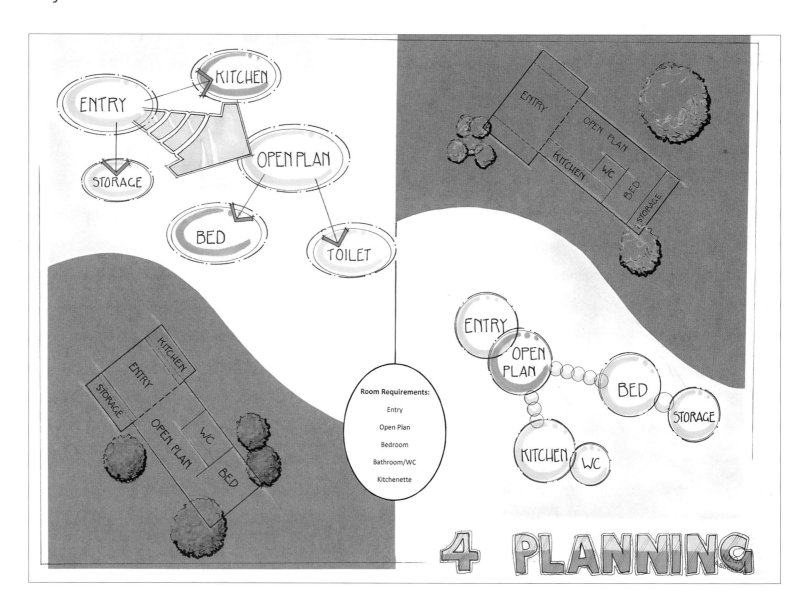

ISBN: 9780170355575

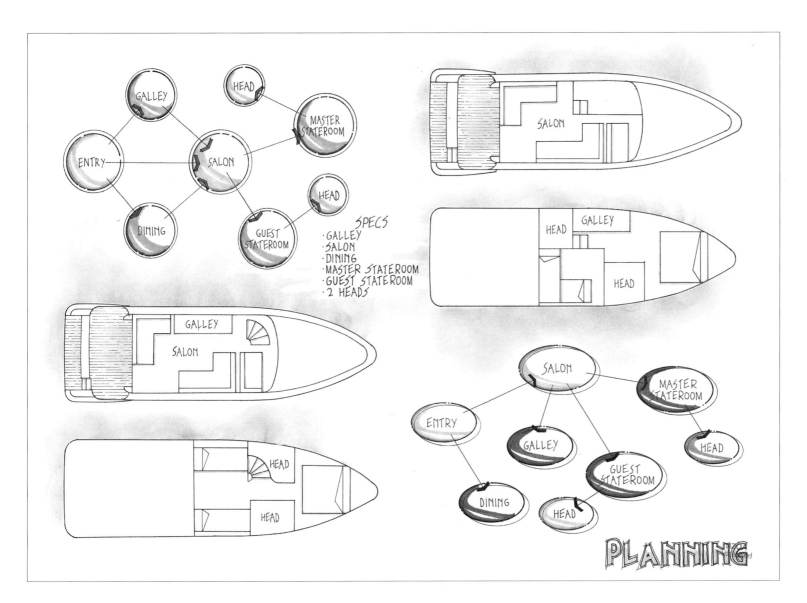

SPECS
· GALLEY
· SALON
· DINING
· MASTER STATEROOM
· GUEST STATEROOM
· 2 HEADS

PLANNING

Getting the curved outline of the yacht and all its interior details in the drawing was the hardest part of this page.

I planned this page again on scrap paper beforehand. I'm glad I did this because I wanted to get a nice balance to it.

I used a circle and an ellipse template for the bubbles, but did them in pencil first to get them the right sizes. Then I rendered them with marker pen and outlined them with black pens. This was the easiest and fun part of the drawing.

I'm glad I learned how to do this in Year 10, because they probably would have turned out messy otherwise.

I wrote down the specs to remember the rooms in the yacht and used blue smudged pastel again for a background to match the other pages.

I took a bit more care doing the title for this page. I like how it turned out.

SITUATION

Traditionally and culturally, Kiwi summer holidays take place in the great outdoors. It's a time to experience the laid-back nature of New Zealand life and travel to hidden corners of our country.

BRIEF: Design a mobile home and a feature for the home.
DURATION: 20 weeks (class time)

REQUIREMENTS   PART ONE: SPATIAL DESIGN

1   Design research. *Cut and pasted among your sketches.*
2   Ideation of the exterior of the home. *Show thumbnails, explorative sketches and alternative ideas. Three A3 pages minimum.*
3   Two bubble diagrams to show planning of the interior.
4   Place notes at any stage, about your research and sketches, to explain thinking and details that are not clear visually. *Discuss: sustainable building materials, energy generation (solar power/heating, etc.), waste management.*
5   Demonstrate in your sketches the effective development of your design ideas, *e.g. explore alternative ideas, not limited to only the brief and its requirements, then refine your chosen ideas informed by aesthetics and function. Support design judgements with qualitative and/or quantitative data gained through research, reflecting your values, tastes and/or views. Where appropriate, use drawings, models, digital modelling, etc.*

6   Draw with instruments, a detailed, scaled third angle orthographic projection of your mobile home design. Show the following:
   - A minimum of three views, fully dimensioned. *Draw the plan as a floor plan to show interior layout, with one elevation either a part section or full section, with cutting plane and cross-hatching.*
   - Hidden detail, cross-hatching, notation and reference lines.
   - A title block with the scale and projection symbol.
   - Correct drawing standards and conventions.

7   Using media of your choice, produce a hand-rendered presentation drawing of your design, in a landscape setting, that effectively communicates its shape and surface qualities.

DESIGN SPECIFICATIONS

The home should:
- be designed with sustainability in mind and accommodate the user in comfort
- be able to be moved when needed — either on wheels or lifted to a new site
- consider both aesthetics and function.
   Function: construction, safety, user friendliness, fit for purpose, ergonomics, etc.
   Aesthetics: style, proportion, finish, harmony, form, etc.

Instrumental drawing

ELEVATIONS AND FLOOR PLAN

These drawings provide another important dimension to an architectural design portfolio by providing accurate, scaled information on sizes, and the layout of the interior as seen in the floor plan.

Although the views do not have to be set out as an orthographic projection, the students featured have done so for assessment purposes, and to satisfy the brief requirements.

Before these drawings can be attempted, classwork exercises must be undertaken, directed at every step by the teacher.

Third angle orthographic projection needs to be introduced in earlier years of study, so that students are not attempting to draw their designs in this format from the outset without any prior knowledge.

Because a more complicated orthographic projection drawing is expected in the product design part of this brief, the drawings featured are simple three or four view layouts.

To satisfy the architectural component, the plan is a floor plan of the home, while the end elevation shows a simple full section from a cutting plane.

SKILLS NEEDED TO PRECEDE THIS PAGE:

Line weights and line standards (2H pencil for all lines).
Title block with scale and projection symbol.
Printing standards (HB pencil).
Correct dimensioning techniques for third angle projection.
Cutting plane and cross-hatching.
Views labelled.
Links to: pages 178–183, 184–185

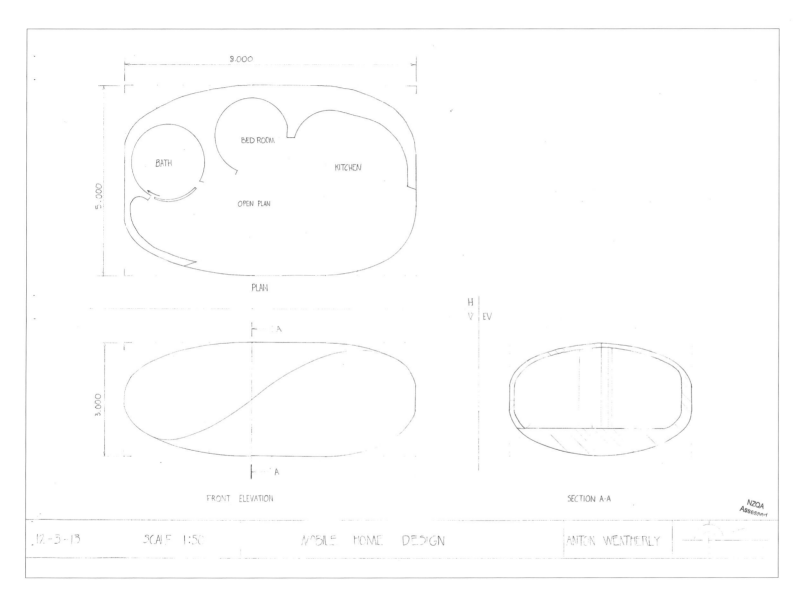

9.000

5.000

BATH

BED ROOM

KITCHEN

OPEN PLAN

PLAN

H
V | EV

3.000

A

A

FRONT ELEVATION

SECTION A-A

NZQA
Assessed

12-3-13   SCALE 1:50   MOBILE HOME DESIGN   ANTON WEATHERLY

Because I knew this drawing was going to be part of my assessment, I made it a traditional orthographic projection. The plan became the floor plan of my design from my bubble diagrams on the previous page.

Instrumental drawings are not my main area of skill. I did not quite get the right hierarchy of lines on the page.

I had to practise this page a few times and even after, I did not get the difference right between construction lines and outlines.

The time-consuming part was getting the right scale to fit the page, then getting the layout right, leaving enough room for dimensions, and cutting plane.

I used a clutch pencil with a 2H lead for the lines and this helped get them better. But I needed to keep sharpening it all the time so this slowed the drawing down.

Using a circle and an ellipse template helped get the curved parts smooth.

Nicole Dealey

I was a bit worried when I first started this drawing that the scale was too small to show enough detail. But it turned out okay in the end.

I planned the page set-out first and I'm glad I did this. It took a bit of time to get all the spaces around the views right.

I drew all the views in light construction first, then did the detail in the plan. This was easy because it was the same layout as my bubble diagram.

Because there is quite a bit of detail, I learned from previous drawings to do all the lines in light construction first so that mistakes can be rubbed out and fixed before doing outlines. I had to do this a few times, especially the front elevation, which was a cross-section.

I used a 2H pencil for all the lines. The hardest part was getting them the right weight; light, medium and dark. I put the labels on last and made guide lines to keep them neat, using an HB pencil.

I had to check the cutting plane a few times because I kept putting the arrows in the wrong direction.

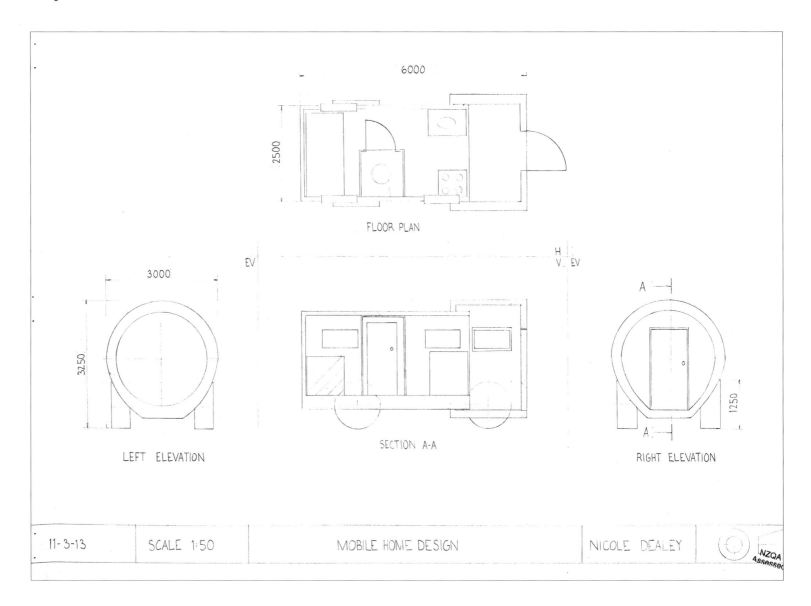

ISBN: 9780170355575   PHOTOCOPYING OF THIS PAGE IS RESTRICTED UNDER LAW.

Mackenzie Farrell

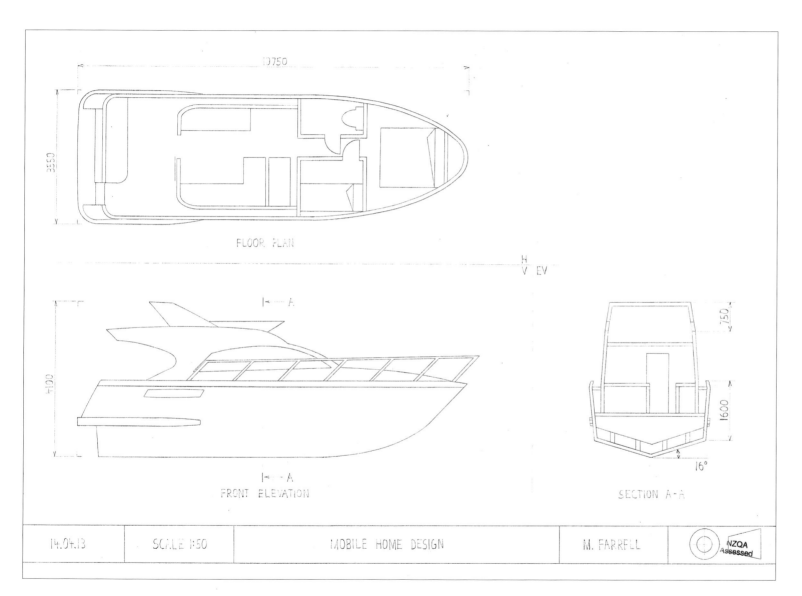

10750

3550

FLOOR PLAN

H
V EV

4100

I← – A

I← – A
FRONT ELEVATION

750

1600

16°

SECTION A-A

| 14.04.13 | SCALE 1:50 | MOBILE HOME DESIGN | M. FARRELL | NZQA Assessed |

This was my first instrumental drawing for the year, so I took a long time to get all the details right. It was good that I had done orthographic drawings in Year 10 and some in class this year as practice first, as my linework was a bit rusty to start with.

I wanted to show two end elevations, but because I needed to show hidden detail, that was going to make the scale too small to see it clearly, so I used only one.

The hardest part was setting the drawings out on the page to give room for dimensions and labels, and to make the page look balanced.

I practised getting the right lines beforehand. I used a clutch pencil with a 2H lead to give me sharp outlines.

After I had the drawing done in construction, I spent time checking that everything was correct before outlining. I found doing medium-weight lines for the reference lines and the cross-hatching the hardest.

I used an ellipse template for the curved parts.

SPATIAL DESIGN

SITUATION
Traditionally and culturally, Kiwi summer holidays take place in the great outdoors. It's a time to experience the laid-back nature of New Zealand life and travel to hidden corners of our country.

BRIEF: Design a mobile home and a feature for the home.
DURATION: 20 weeks (class time)

REQUIREMENTS   PART ONE: SPATIAL DESIGN
1. Design research. *Cut and pasted among your sketches.*
2. Ideation of the exterior of the home. *Show thumbnails, explorative sketches and alternative ideas. Three A3 pages minimum.*
3. Two bubble diagrams to show planning of the interior.
4. Place notes at any stage, about your research and sketches, to explain thinking and details that are not clear visually. *Discuss: sustainable building materials, energy generation (solar power/heating, etc.), waste management.*
5. Demonstrate in your sketches the effective development of your design ideas, *e.g. explore alternative ideas, not limited to only the brief and its requirements, then refine your chosen ideas informed by aesthetics and function. Support design judgements with qualitative and/or quantitative data gained through research, reflecting your values, tastes and/or views. Where appropriate, use drawings, models, digital modelling, etc.*
6. Draw with instruments, a detailed, scaled third angle orthographic projection of your mobile home design. Show the following:
   - A minimum of three views, fully dimensioned. *Draw the plan as a floor plan to show interior layout, with one elevation either a part section or full section, with cutting plane and cross-hatching.*
   - Hidden detail, cross-hatching, notation and reference lines.
   - A title block with the scale and projection symbol.
   - Correct drawing standards and conventions.

7. Using media of your choice, produce a hand-rendered presentation drawing of your design, in a landscape setting, that effectively communicates its shape and surface qualities.

DESIGN SPECIFICATIONS
The home should:
- be designed with sustainability in mind and accommodate the user in comfort
- be able to be moved when needed — either on wheels or lifted to a new site
- consider both aesthetics and function.
  Function: construction, safety, user friendliness, fit for purpose, ergonomics, etc.
  Aesthetics: style, proportion, finish, harmony, form, etc.

Presentation drawing

Now it's time to bring out the big guns and show the chosen design to advantage. The drawings have been sketched freehand until the correct proportions and viewpoint were achieved.

Then several photocopies were made on thicker paper (cartridge paper), to allow marker pens to be used. One photocopy was preserved as the master, the others were used for practice rendering of the parts. Once satisfied, these were cut out and pasted onto the master.

To precede this final drawing, and to effectively communicate shape and surface qualities, students must be taught rendering techniques using a class exercise. As all designs are different, students must be practised in using a range of media to gain an understanding of the best type to use for their design. Because Anton's design is curved, he has chosen to use an airbrush. The water and drop-down side have been rendered in chalk pastel. An eraser has been used to create shine on the drop-down surface.

Again, these simple techniques must be taught and practised — they are not a given.

The finished rendering has been cut out and pasted onto its background. Careful consideration and choice of background enhances the final presentation.

The addition of a human person also gives a sense of scale.

Note the subtle use of colour, which once again maintains the same palette as all preceding pages.

 SKILLS NEEDED TO PRECEDE THIS PAGE:

Perspective drawing techniques.
Viewpoint, backgrounds.
Crating (proportions) and line hierarchy.
Rendering techniques — light source, tonal changes, shape and surface texture.
Colour palette considered.
Media technique: marker pens, black fineline pens, pastel pencils.
Links to: pages 30–34, 189–191

ISBN: 9780170355575

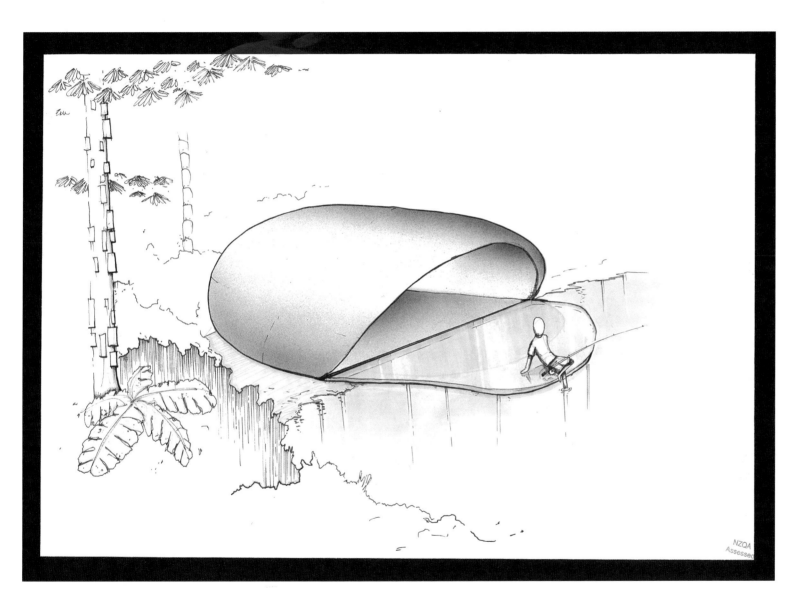

I thought a lot about how best to show my design and how best to render it.

I experimented first with a number of different layouts and viewpoints, but decided on a more isometric view.

After experimenting with markers for rendering, I could not get the look I wanted, so decided on the airbrush. It was good that I had one in my classroom and had used it before. It was ideal for showing curved surfaces.

I didn't have to mask the drawing beforehand, as I simply cut it out after airbrushing, then glued it to the landscape background.

I spent a lot of time looking for a background as I wanted to show my design in a landscape setting. I sketched simple 3D plant forms and outlined them with a black fineline pen. The hardest part was rendering the water. I had to attempt this several times to get it right, and although it still isn't realistic enough, I am happy with the end result.

Nicole Dealey

I wanted this page to really show my design off, so I took a lot of time over it.

I sketched my final design, then made four photocopies of it. I practised on the photocopies to get the rendering right. I needed to make some more because I used them all up getting the reflection on the windows to look right.

I took two attempts to draw and render the wheels, which I still think are not quite right.

When I had finished rendering the home, I cut it out and made the background to paste it on. I did a silhouette of a person to see the scale better.

I kept all the parts in the background simple, then filled the shapes in with a black marker pen.

If I was to do this again, I would change the person and the large black shadow by making them a bit lighter.

Mackenzie Farrell

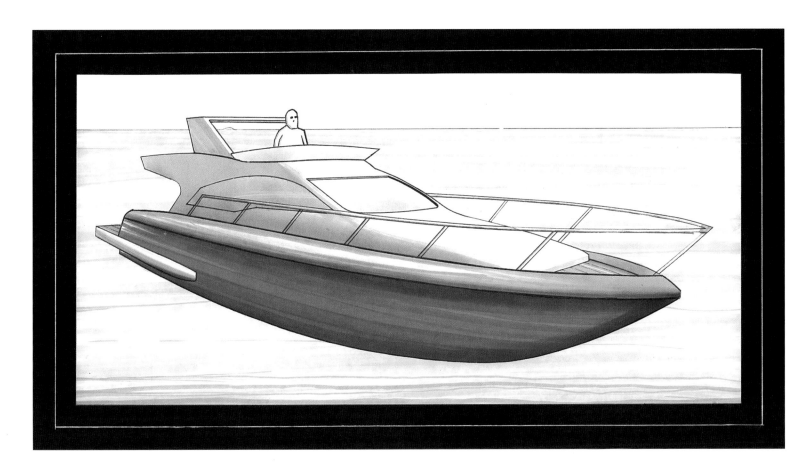

I knew I had to get the hull of the yacht correct as it is the largest and focal part of the drawing.

It took a long time to decide on the viewpoint and how to render the yacht.

I did many practice pages first until I had some techniques that achieved the desired result.

I re-sketched the drawing from my design pages, then made photocopies to practise on.

I used mainly marker pens for the rendering. The hardest part was rendering the aluminium parts with different tones of grey marker.

I was running out of time on this page, but in the end it turned out okay. If I was to do it again I would use a more interesting background. It was hard deciding on one that was detailed but did not detract from the main focus of the drawing.

SITUATION

Traditionally and culturally, Kiwi summer holidays take place in the great outdoors. It's a time to experience the laid-back nature of New Zealand life and travel to hidden corners of our country.

BRIEF: Design a mobile home and a feature for the home.
DURATION: 20 weeks (class time)

REQUIREMENTS   **PART TWO: PRODUCT DESIGN**

1 Design research. *Cut and pasted among your sketches.*
2 Ideation of the feature. *Show: thumbnails, explorative sketches, thinking and technical sketches. Three A3 pages minimum.*
3 The feature must have a compound curve, circular parts and/or angled parts that can be plotted in instrumental drawings.
4 Place notes at any stage, about your research and sketches, to explain thinking and details that are not clear visually.
5 Demonstrate in your sketches the effective development of your design ideas, *e.g. explore alternative ideas, not limited to only the brief and its requirements, then refine your chosen ideas informed by aesthetics and function. Support design judgements with qualitative and/or quantitative data gained through research, reflecting your values, tastes and/or views. Where appropriate, use drawings, models, digital modelling, etc.*

6 Draw with instruments, a detailed, scaled third angle orthographic projection of your feature design. Show the following:
   - A minimum of three views, fully dimensioned.
   - Section one elevation to show construction detail.
   - A parts list with numbered components about the views.
   - Hidden detail, cross-hatching, notation and reference lines.
   - A title block with the scale and projection symbol.
   - Correct drawing standards and conventions.
7 Draw with instruments, a scaled paraline drawing of your design (oblique or isometric) and an exploded paraline drawing of a part of the design, *i.e. the joining method found in the construction of the part. Show a parts list and all constructions clearly.*
8 Using media of your choice, produce a hand-rendered presentation drawing of your design to show shape and surface qualities.

DESIGN SPECIFICATIONS

- The feature could be: *a bench/sink unit, bunks, table, outdoor seat, GPS, shelving, outdoor barbecue, etc.*
- The feature must: *consider human factors and be designed for either the interior or the exterior of the home.*
- It should consider both aesthetics and function.
  Function: construction, safety, user friendliness, fitness for purpose, ergonomics, etc.
  Aesthetics: style, proportion, finish, harmony, form, etc.

Ideation

THUMBNAILS AND EXPLORATIVE SKETCHES

The second part of this brief is for the design of a product for the mobile home designed in part one. Requirements for ideation can now be more detailed, with a focus on construction, materials, sustainability and human factors.

This activity provides the student with a wider range of learning experiences, and more accurate assessment opportunites, by combining ideation sets from parts one and two.

The development of a sophisticated sketching style should continue to be encouraged at this level.

Although not necessary until the technical sketching page, the introduction of a 2D sectioned view, as seen on Anton's page, confirms student thinking about construction that will inform the design later on.

As in the first ideation set, careful attention has been given to layout, viewpoints, ensuring a mix of 2D and 3D sketches, colour palette and page background. Intentionally minimalist rendering provides adequate information about form.

Shadows and backgrounds provide visual interest to the page, while good use of black ballpoint pen on craft paper again enhances the thumbnails.

Titles should match those in the first set but, more importantly, indicate what the student has designed.

SKILLS NEEDED TO PRECEDE THIS PAGE:

A planning sheet, freehand design sketching. Thumbnails — use a black ballpoint pen on craft paper.

Exploratives (2D and 3D), viewpoints, shadows, crating (proportions) and line hierarchy.

Rendering techniques — light source, tonal changes, form.

Research and simple design notes that explain thinking not clear visually.

Media technique: marker pens, black fineline pens, pastel pencils.

Links to: pages 75–81

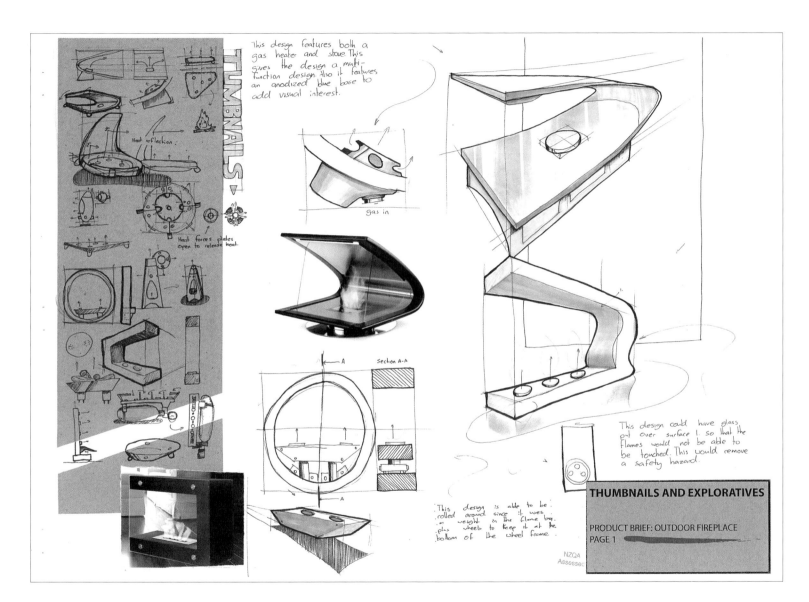

This design features both a gas heater and stove. This gives the design a multi-function design. Also it features an anodized blue base to add visual interest.

gas in

Heat reflection.

Heat forces plates open to release heat.

Section A-A

This design could have glass put over surface 1. so that the flames would not be able to be touched. This would remove a safety hazard.

This design is able to be rolled around since it uses weight in the flame box plus wheels to keep it at the bottom of the wheel frame.

THUMBNAILS AND EXPLORATIVES

PRODUCT BRIEF: OUTDOOR FIREPLACE
PAGE 1

NZQA
Assessed

I found this page a lot easier than the first one, mainly because this was a repeat but with a different design.

To match the other pages, I again tried a styley look but with steeper angles to some of the sketches for better effect.

I thought at first that my thumbnails were too small to show details, then had to keep reminding myself that they should be quick sketches of my first ideas and don't need to show as much detail as I did.

I used 2D and 3D sketches, which I feel give the page a good look and explain things clearly.

I kept everything simple and used minimal rendering, just shading in the surfaces away from the light. I think this works well and makes the rendering quick.

Nicole Dealey

Although I wasn't going to do a practice page first, I found that I had to stop and do one because the drawings weren't looking as well placed as I wanted.

Getting different viewpoints and the right porportions for the crates took a while.

I used the same techniques for drawing and rendering as my first page, so it didn't take as long for this one.

I would do some more thumbnails if I was to do the page again, but I like the way they turned out.

I kept the rendering greyscale again, using markers, and used a sandstone marker for the wood parts.

I tried to get different viewpoints on all the sketches and used thick lines to make them clearer on the page.

I did the same title block on a computer to match my other pages, and printed it onto the page before I started drawing.

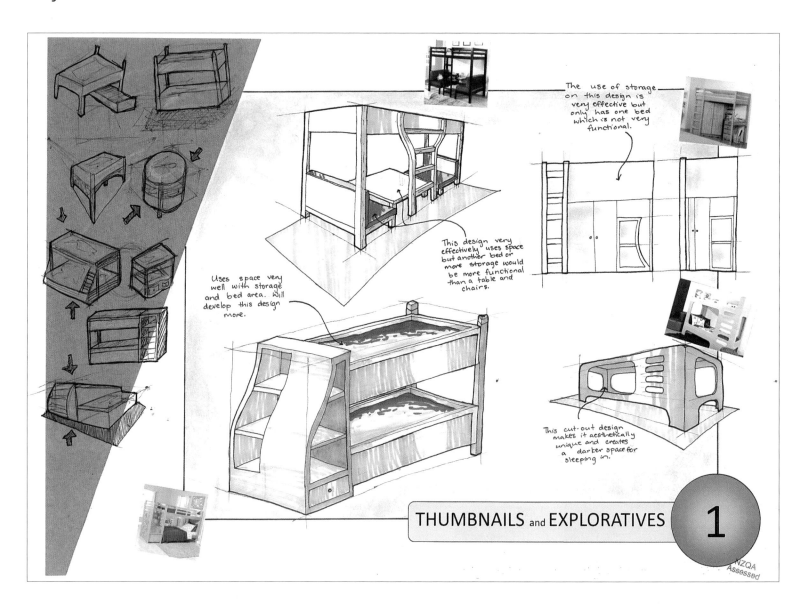

The use of storage on this design is very effective but only has one bed which is not very functional.

This design very effectively uses space but another bed or more storage would be more functional than a table and chairs.

Uses space very well with storage and bed area. Will develop this design more.

This cut-out design makes it aesthetically unique and creates a darker space for sleeping in.

THUMBNAILS and EXPLORATIVES ①

NZQA Assessed

ISBN: 9780170355575

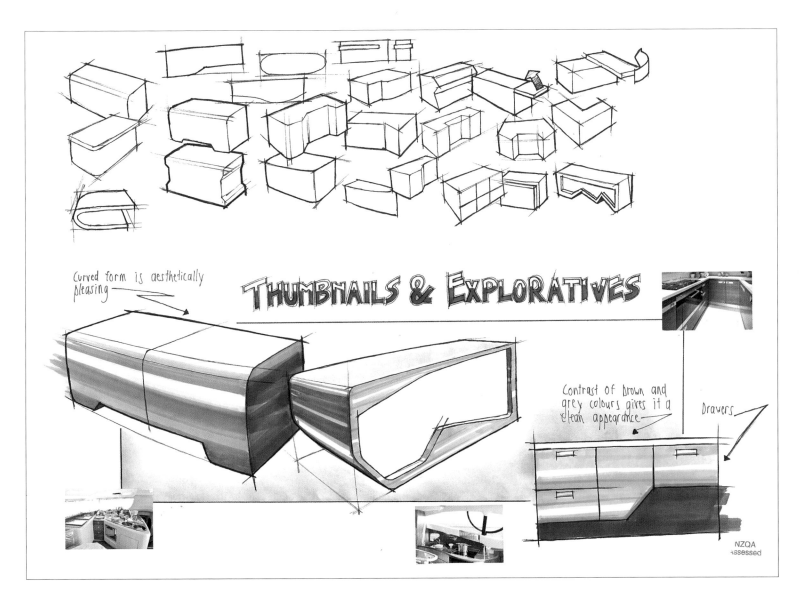

THUMBNAILS & EXPLORATIVES

Curved form is aesthetically pleasing

Contrast of brown and grey colours gives it a clean appearance

Drawers

NZQA
Assessed

Even though this is quite a simple-looking page, like my thumbnail page for the yacht, I planned all parts of the page beforehand.

I kept the thumbnails across the top to match the first page, and also give more space for longer designs for a galley. I was going to use brown paper underneath, but decided to do them on the white backgrounds to make them stand out. I kept them line drawings to better show the outline shapes I was thinking of.

Even though I only showed three explorative designs, I would show more if I was to do the page again.

I used the same rendering media of marker pen and light-blue smudged pastel pencil for the background.

Research helped a lot with the designs for this part of the assignment.

SITUATION

Traditionally and culturally, Kiwi summer holidays take place in the great outdoors. It's a time to experience the laid-back nature of New Zealand life and travel to hidden corners of our country.

BRIEF: Design a mobile home and a feature for the home.
DURATION: 20 weeks (class time)

REQUIREMENTS

PART TWO: PRODUCT DESIGN

1   Design research. *Cut and pasted among your sketches.*
2   Ideation of the feature. *Show: thumbnails, explorative sketches, thinking and technical sketches. Three A3 pages minimum.*
3   The feature must have a compound curve, circular parts and/or angled parts that can be plotted in instrumental drawings.
4   Place notes at any stage, about your research and sketches, to explain thinking and details that are not clear visually.
5   Demonstrate in your sketches the effective development of your design ideas, *e.g. explore alternative ideas, not limited to only the brief and its requirements, then refine your chosen ideas informed by aesthetics and function. Support design judgements with qualitative and/or quantitative data gained through research, reflecting your values, tastes and/or views. Where appropriate, use drawings, models, digital modelling, etc.*

6   Draw with instruments, a detailed, scaled third angle orthographic projection of your feature design. Show the following:
    - A minimum of three views, fully dimensioned.
    - Section one elevation to show construction detail.
    - A parts list with numbered components about the views.
    - Hidden detail, cross-hatching, notation and reference lines.
    - A title block with the scale and projection symbol.
    - Correct drawing standards and conventions.
7   Draw with instruments, a scaled paraline drawing of your design (oblique or isometric) and an exploded paraline drawing of a part of the design, *i.e. the joining method found in the construction of the part. Show a parts list and all constructions clearly.*
8   Using media of your choice, produce a hand-rendered presentation drawing of your design to show shape and surface qualities.

DESIGN SPECIFICATIONS

- The feature could be: *a bench/sink unit, bunks, table, outdoor seat, GPS, shelving, outdoor barbecue, etc.*
- The feature must: *consider human factors and be designed for either the interior or the exterior of the home.*
- It should consider both aesthetics and function.
  Function: construction, safety, user friendliness, fitness for purpose, ergonomics, etc.
  Aesthetics: style, proportion, finish, harmony, form, etc.

Ideation (cont.)

ALTERNATIVES AND THINKING SKETCHES

Again, after initial exploration of ideas, students must be given the opportunity to take their thinking further and explore alternatives that could better inform the design.

The organic shapes provided by nature, as seen in all students' research, clearly have an influence on the design.

A lesson on bio-mimickry provokes thinking in the right direction.

Note the continuation of the same sketching style in 2D and 3D, and the dramatic use of viewpoint to highlight shape and form.

Particularly effective is the view looking from beneath. Crating is clearly defined, offering correct proportion, while thick and thin lines (line hierarchy) make the sketches 'pop' on the page. Sketching on craft paper adds to the 'work in progress' feel to the page.

This technique is also useful if the student wants to cover up a 'mistake' or an idea or viewpoint that is not working.

Once again, minimalist marker rendering conveys tone and shape effectively. Plenty of white space allows sketches to 'breathe' on the page.

SKILLS NEEDED TO PRECEDE THIS PAGE:

A planning sheet.

Freehand design sketching in 2D and 3D with clearly seen crating (proportions) and line hierarchy.

Viewpoints are considered for visual interest and to better show the ideas.

Arrows indicate movement and/or function.

Rendering techniques — minimalist, indicating the light source, tonal changes and form.

Colour palette — soft tones.

Links to: pages 75–81

ISBN: 9780170355575   PHOTOCOPYING OF THIS PAGE IS RESTRICTED UNDER LAW.

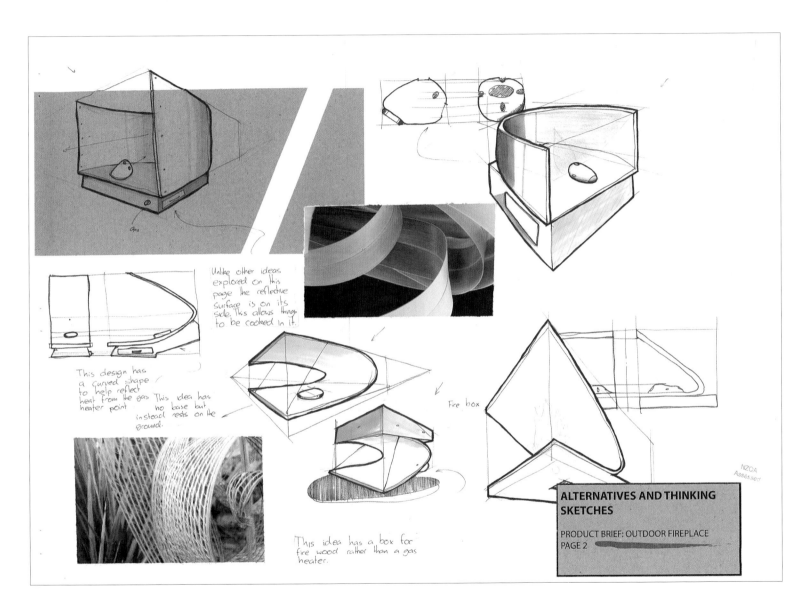

Gas

Unlike other ideas explored on this page the reflective surface is on its side. This allows things to be cooked in it.

This design has a curved shape to help reflect heat from the gas heater point. This idea has no base but instead rests on the ground.

Fire box

This idea has a box for fire wood rather than a gas heater.

ALTERNATIVES AND THINKING SKETCHES

PRODUCT BRIEF: OUTDOOR FIREPLACE
PAGE 2

NZQA Assessed

I like this page, especially my sketches and how some of them overlap each other.

Once I knew what I was going to design for the mobile home, by getting some alternative research, especially finding the curves on the flax leaves, gave me more of an idea of the final shape.

I used all the techniques on my other pages so that the look and rendering matched.

I also used a planning sheet to start with to make sure that all the things I wanted to show could be easily seen.

The only thing that I was not happy with was the position of the title block. I only remembered about it after most of the page was finished, so I had to paste it over the top of my best sketch.

Nicole Dealey

It was difficult at first thinking of alternatives because I already knew what my design would look like. When I found pictures of natural things such as the spider web, insect and light coming through an opening, I got more ideas of how to use them in the design.

All the views are drawn in perspective to keep a realistic look.

I planned the page layout again to make sure of the right look and used brown paper again to match my other pages.

For the rendering I used grey marker pen for the main parts of the bunks and sandstone for wood.

On this page I decided on my chosen design, which was the one based on the curve of the insect eyes. I like the way this has added an interesting feature to the design which I wouldn't have thought of.

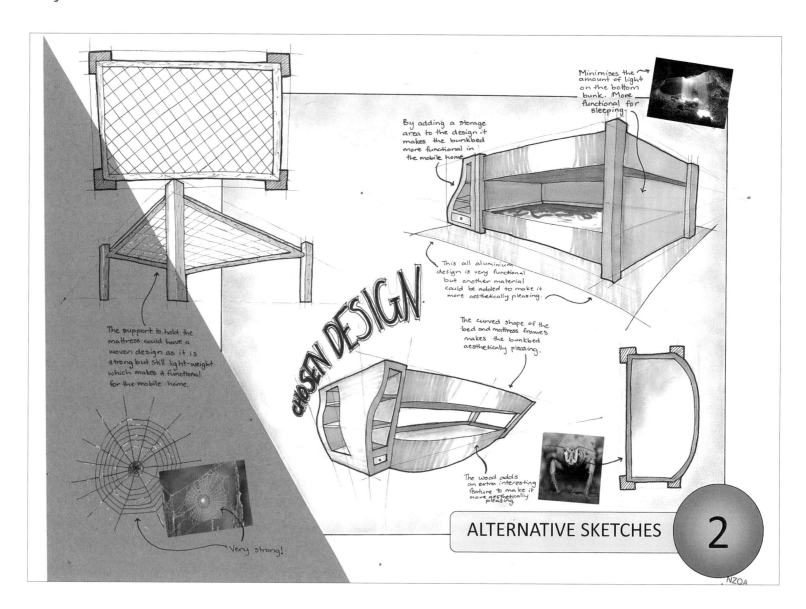

Minimises the amount of light on the bottom bunk. More functional for sleeping.

By adding a storage area to the design it makes the bunkbed more functional in the mobile home.

This all aluminium design is very functional but another material could be added to make it more aesthetically pleasing.

The curved shape of the bed and mattress frames makes the bunkbed aesthetically pleasing.

CHOSEN DESIGN

The support to hold the mattress could have a woven design as it is strong but still light-weight which makes it functional for the mobile home.

The wood adds an extra interesting feature to make it more aesthetically pleasing.

Very strong!

ALTERNATIVE SKETCHES   2

NZQA

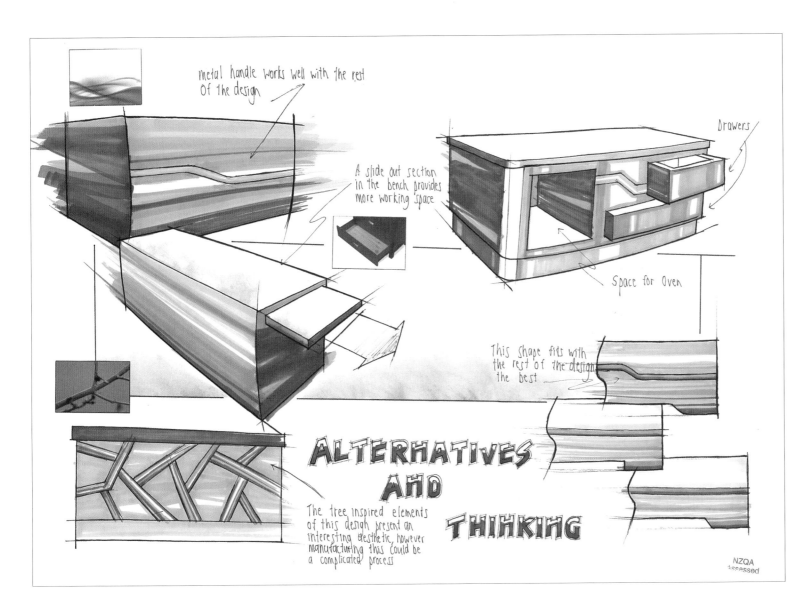

metal handle works well with the rest of the design

A slide out section in the bench provides more working space

Drawers

Space for Oven

This shape fits with the rest of the design the best

ALTERNATIVES AND THINKING

The tree inspired elements of this design present an interesting aesthetic, however manufacturing this could be a complicated process

NZQA
assessed

I kept this page simple and focused on a design element in the form of a zigzig through the side to give a more stylish look to the galley design.

I found the look I wanted that was inspired by tree branches. The way they interlock, and their shapes, added a nice touch to the side.

I did a quick layout plan for the page first, to make sure the drawings fit and looked pleasing.

I used isometric and perspective for the 3Ds and different viewpoints to make the look more exciting.

After rendering my other pages, I found rendering this page was easy. I knew how to render the materials with marker pens and am really happy with the way it turned out.

I kept the colour palette and style of lettering for the title the same as my other pages.

SITUATION

Traditionally and culturally, Kiwi summer holidays take place in the great outdoors. It's a time to experience the laid-back nature of New Zealand life and travel to hidden corners of our country.

BRIEF: Design a mobile home and a feature for the home.
DURATION: 20 weeks (class time)

REQUIREMENTS   PART TWO: PRODUCT DESIGN

1. Design research. *Cut and pasted among your sketches.*
2. Ideation of the feature. *Show: thumbnails, explorative sketches, thinking and technical sketches. Three A3 pages minimum.*
3. The feature must have a compound curve, circular parts and/or angled parts that can be plotted in instrumental drawings.
4. Place notes at any stage, about your research and sketches, to explain thinking and details that are not clear visually.
5. Demonstrate in your sketches the effective development of your design ideas, *e.g. explore alternative ideas, not limited to only the brief and its requirements, then refine your chosen ideas informed by aesthetics and function. Support design judgements with qualitative and/or quantitative data gained through research, reflecting your values, tastes and/or views. Where appropriate, use drawings, models, digital modelling, etc.*

6. Draw with instruments, a detailed, scaled third angle orthographic projection of your feature design. Show the following:
   - A minimum of three views, fully dimensioned.
   - Section one elevation to show construction detail.
   - A parts list with numbered components about the views.
   - Hidden detail, cross-hatching, notation and reference lines.
   - A title block with the scale and projection symbol.
   - Correct drawing standards and conventions.
7. Draw with instruments, a scaled paraline drawing of your design (oblique or isometric) and an exploded paraline drawing of a part of the design, *i.e. the joining method found in the construction of the part. Show a parts list and all constructions clearly.*
8. Using media of your choice, produce a hand-rendered presentation drawing of your design to show shape and surface qualities.

DESIGN SPECIFICATIONS

- The feature could be: *a bench/sink unit, bunks, table, outdoor seat, GPS, shelving, outdoor barbecue, etc.*
- The feature must: *consider human factors and be designed for either the interior or the exterior of the home.*
- It should consider both aesthetics and function.
  Function: construction, safety, user friendliness, fitness for purpose, ergonomics, etc.
  Aesthetics: style, proportion, finish, harmony, form, etc.

Ideation (cont.)

TECHNICAL SKETCHES

Now is the time to begin thinking about construction details of the design that could allow it to be made. Technical sketches need to be a mix of 3D exploded isometric and 2D sectioned views of parts.

Anton has showed this particularly well, confirming not only his understanding of how his design could be made but also his knowledge of sketching technique.

Note the size of each sketch. A larger sketch makes understanding clearer, particularly effective in Anton's cutaway isometric drawing, and in his 2D sectioned elevation.

Good use of line hierarchy (thick and thin lines) on all students' sketches provide visual impact. Rendering is mostly applied to highlight tone and surface shape.

The use of craft paper provides visual interest, while plenty of white space allows his sketches to 'breathe'.

⊙ SKILLS NEEDED TO PRECEDE THIS PAGE:

A planning sheet used.

Freehand design sketching in 2D and 3D and line hierarchy.

Viewpoints are considered for visual interest and to better show the ideas.

Correct 'exploded' techniques. The parts are pulled apart in the direction in which they would be assembled.

Correct cross-hatching technique from a cutting plane.

Simple notes explain thinking not clear visually.

Rendering techniques — minimalist, indicating the light source, tonal changes and form.

Colour palette — soft grey tones, light blues and black.

Links to: pages 75–81

ISBN: 9780170355575

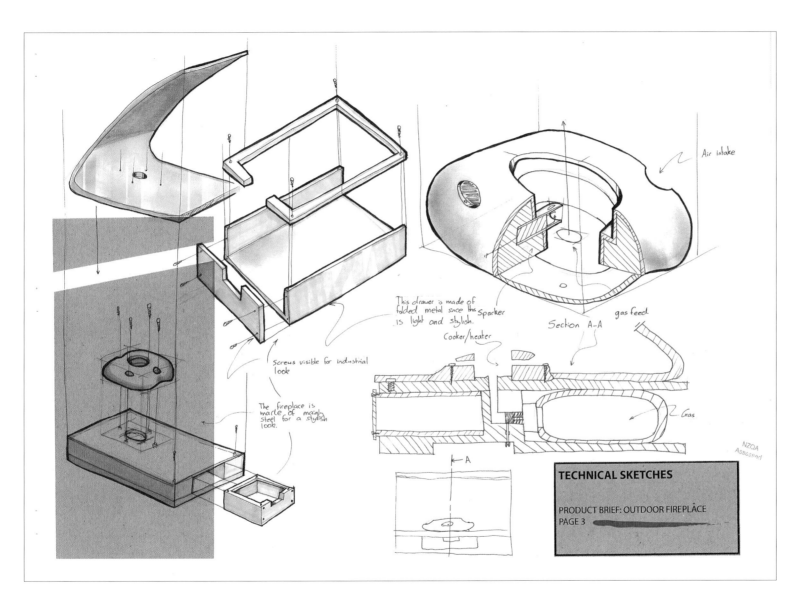

Air intake

This drawer is made of folded metal since this is light and stylish.

Sparker

Screws visible for industrial look

Cooker/heater

Section A–A

gas feed

The fireplace is made of mainly steel for a stylish look.

Gas

← A

NZQA Assessed

TECHNICAL SKETCHES

PRODUCT BRIEF: OUTDOOR FIREPLACE
PAGE 3

After my first technical page, which I found hard, this one was a lot easier. I knew how I wanted to make the design, so I just had to decide on the best types of sketches.

I made all the sketches quite large, so all the detail could be seen easily. I learned this from my first page where some were too small.

I used 3D exploded and cutaway views, and 2D sectioned views. I knew about these types of drawings from drawings I had done in class earlier, and am really pleased with the way they turned out.

I kept the rendering simple again, just shading the darker surfaces using a light-grey marker pen. The black thick and thin outlines make the sketches look more 3D.

Nicole Dealey

I had to do a lot of research before doing this page. I was not sure how the parts would be made, so I looked on the internet and in books I had in my classroom.

Because I did not have much room on the page, I decided to show only a few of the main parts instead of the whole design.

I'm glad I had done exploded sketches in Year 10. This helped a lot, even though the drawing of detail A2 is shown pulled apart in the wrong direction. I only noticed this after I had finished but ran out of time to fix it.

I planned the page layout first, and although it is quite crowded, the arrows showing function make it flow better.

I kept the rendering simple using a marker pen and a black pen for the outlines.

The brown paper down the side makes the page match my others.

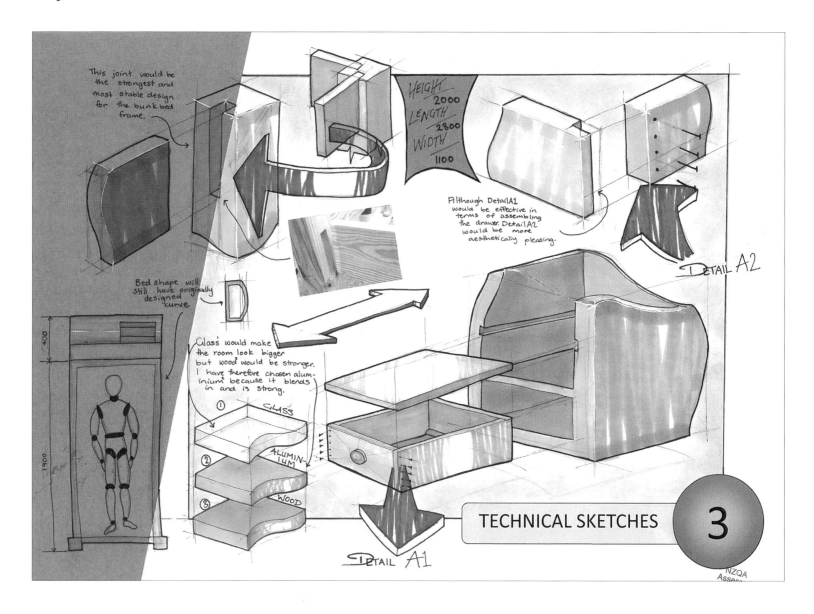

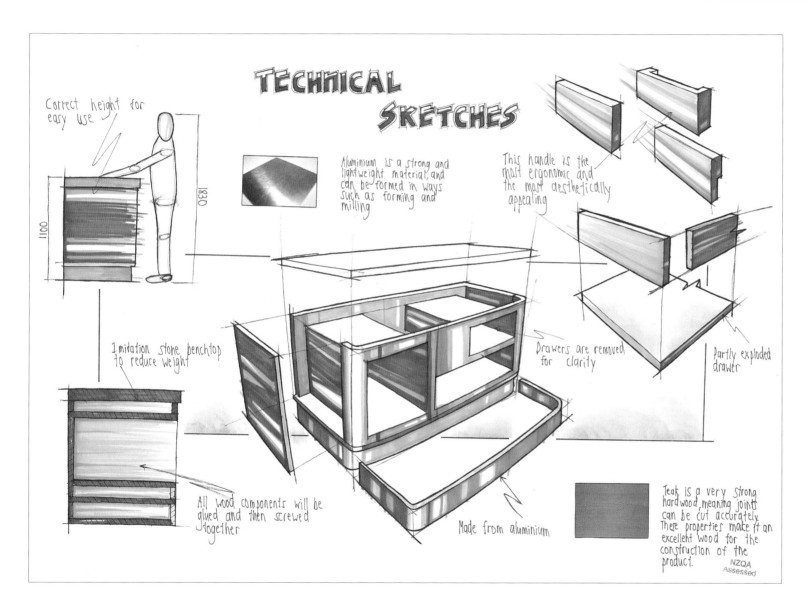

TECHNICAL SKETCHES

Correct height for easy use

1100

1830

Aluminium is a strong and lightweight material, and can be formed in ways such as forming and milling

This handle is the most ergonomic and the most aesthetically appealing

Imitation stone benchtop to reduce weight

Drawers are removed for clarity

Partly exploded drawer

All wood components will be glued and then screwed together

Made from aluminium

Teak is a very strong hardwood, meaning joints can be cut accurately. These properties make it an excellent wood for the construction of the product.

NZQA Assessed

After I designed this part, I found that it was going to be fairly complex to show it exploded. I looked at the galley in my parents' boat, and that helped, along with doing some research, to get the construction of it right.

To begin with, I planned where the dominant image would be on the page, and what angle it was going to be viewed from.

Learning how to draw in three-point perspective, which I had done in Year 10, allowed me to present my ideas with a greater degree of realism than isometric or one-point perspective.

I did the drawings in pencil first, and deliberately used steep angles on some to make them more stylish.

The rendering didn't take as long as I thought it would. I used a range of grey and brown coloured markers to get the look of shiny wood and metal. But I practised on some photocopies first to make sure it was right.

SITUATION

Traditionally and culturally, Kiwi summer holidays take place in the great outdoors. It's a time to experience the laid-back nature of New Zealand life and travel to hidden corners of our country.

BRIEF: Design a mobile home and a feature for the home.
DURATION: 20 weeks (class time)

REQUIREMENTS

PART TWO: PRODUCT DESIGN

1   Design research. *Cut and pasted among your sketches.*
2   Ideation of the feature. *Show: thumbnails, explorative sketches, thinking and technical sketches. Three A3 pages minimum.*
3   The feature must have a compound curve, circular parts and/or angled parts that can be plotted in instrumental drawings.
4   Place notes at any stage, about your research and sketches, to explain thinking and details that are not clear visually.
5   Demonstrate in your sketches the effective development of your design ideas, *e.g. explore alternative ideas, not limited to only the brief and its requirements, then refine your chosen ideas informed by aesthetics and function. Support design judgements with qualitative and/or quantitative data gained through research, reflecting your values, tastes and/or views. Where appropriate, use drawings, models, digital modelling, etc.*

6   Draw with instruments, a detailed, scaled third angle orthographic projection of your feature design. Show the following:
   ▪   A minimum of three views, fully dimensioned.
   ▪   Section one elevation to show construction detail.
   ▪   A parts list with numbered components about the views.
   ▪   Hidden detail, cross-hatching, notation and reference lines.
   ▪   A title block with the scale and projection symbol.
   ▪   Correct drawing standards and conventions.
7   Draw with instruments, a scaled paraline drawing of your design (oblique or isometric) and an exploded paraline drawing of a part of the design, *i.e. the joining method found in the construction of the part. Show a parts list and all constructions clearly.*
8   Using media of your choice, produce a hand-rendered presentation drawing of your design to show shape and surface qualities.

DESIGN SPECIFICATIONS

▪   The feature could be: *a bench/sink unit, bunks, table, outdoor seat, GPS, shelving, outdoor barbecue, etc.*
▪   The feature must: *consider human factors and be designed for either the interior or the exterior of the home.*
▪   It should consider both aesthetics and function.
   Function: construction, safety, user friendliness, fitness for purpose, ergonomics, etc.
   Aesthetics: style, proportion, finish, harmony, form, etc.

Ideation (cont.)

MATERIALS AND HUMAN FACTORS

Here is another most important stage in the ideation process, particularly for product design. Now is the time to think about what the design could be made from, and to 'fine tune' it to fit the human form.

This page shows Anton's exploration of the materials that could be used to reflect heat from the fireplace, the sizes of the fireplace (important for the scaled instrumental drawings to follow) and the hand grip for picking it up and moving it to another location.

Note the continuing use of 2D and 3D sketches, line hierarchy (thick and thin lines) that provides the sketches with visual impact, and the use of crating to achieve the right proportion of each sketch. Research is ongoing, continuously informing ideas.

Again, rendering is minimalist, to highlight tone, surface shape and shadows. The use of craft paper continues to provide visual interest to the page. Lots of white space allows his sketches to 'breathe' on the page.

SKILLS NEEDED TO PRECEDE THIS PAGE:

A planning sheet.

Dimensions are indicated for instrumental drawings.

Ongoing research better informs decisions.

Freehand design sketching in 2D and 3D with clearly seen crating (proportions) and line hierarchy.

Viewpoints are considered for visual interest and to better show the ideas.

Rendering techniques — minimalist, indicating the light source, tonal changes and form.

Colour palette — soft grey tones, light blues and black.

Links to: pages 75–81

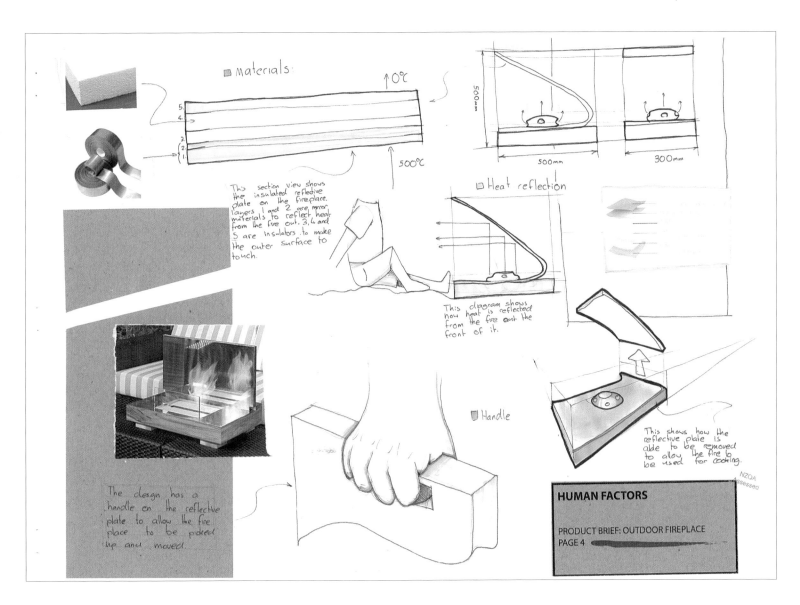

Materials:

↑0°C

500°C

This section view shows the insulated reflective plate on the fireplace. layers 1 and 2 are mirror materials to reflect heat from the fire out. 3, 4 and 5 are insulators to make the outer surface to touch.

500mm

500mm

300mm

Heat reflection

This diagram shows how heat is reflected from the fire out the front of it.

Handle

This shows how the reflective plate is able to be removed to allow the fire to be used for cooking.

The design has a handle on the reflective plate to allow the fire place to be picked up and moved.

NZQA Assessed

HUMAN FACTORS

PRODUCT BRIEF: OUTDOOR FIREPLACE
PAGE 4

To follow on from my technical sketches page, I wanted to look at how my design would suit people who would use it.

I concentrated on the main things — the hand grip for carrying the fireplace, materials it could be made from and how the heat would be reflected.

I had to do a bit of research to find out this information.

I kept my sketches simple, using 2D and 3D, as well as keeping the rendering minimal to match my other pages.

If I was to do this page again, I would have shown more detail. But I was running out of time. I am happy with this page though.

SITUATION

Traditionally and culturally, Kiwi summer holidays take place in the great outdoors. It's a time to experience the laid-back nature of New Zealand life and travel to hidden corners of our country.

BRIEF: Design a mobile home and a feature for the home.
DURATION: 20 weeks (class time)

REQUIREMENTS   **PART TWO: PRODUCT DESIGN**

1   Design research. *Cut and pasted among your sketches.*
2   Ideation of the feature. *Show: thumbnails, explorative sketches, thinking and technical sketches. Three A3 pages minimum.*
3   The feature must have a compound curve, circular parts and/or angled parts that can be plotted in instrumental drawings.
4   Place notes at any stage, about your research and sketches, to explain thinking and details that are not clear visually.
5   Demonstrate in your sketches the effective development of your design ideas, *e.g. explore alternative ideas, not limited to only the brief and its requirements, then refine your chosen ideas informed by aesthetics and function. Support design judgements with qualitative and/or quantitative data gained through research, reflecting your values, tastes and/or views. Where appropriate, use drawings, models, digital modelling, etc.*

6   Draw with instruments, a detailed, scaled third angle orthographic projection of your feature design. Show the following:
    ▪ A minimum of three views, fully dimensioned.
    ▪ Section one elevation to show construction detail.
    ▪ A parts list with numbered components about the views.
    ▪ Hidden detail, cross-hatching, notation and reference lines.
    ▪ A title block with the scale and projection symbol.
    ▪ Correct drawing standards and conventions.

7   Draw with instruments, a scaled paraline drawing of your design (oblique or isometric) and an exploded paraline drawing of a part of the design, *i.e. the joining method found in the construction of the part. Show a parts list and all constructions clearly.*
8   Using media of your choice, produce a hand-rendered presentation drawing of your design to show shape and surface qualities.

DESIGN SPECIFICATIONS

▪ The feature could be: *a bench/sink unit, bunks, table, outdoor seat, GPS, shelving, outdoor barbecue, etc.*
▪ The feature must: *consider human factors and be designed for either the interior or the exterior of the home.*
▪ It should consider both aesthetics and function.
   Function: construction, safety, user friendliness, fitness for purpose, ergonomics, etc.
   Aesthetics: style, proportion, finish, harmony, form, etc.

Instrumental drawing

ORTHOGRAPHIC PROJECTION

Okay, so we all know that instrumental drawings are shunned by many, particularly orthographic projection. But this is a shame. As seen in the featured drawings, they offer important, accurate, scaled information on sizes and construction that could allow the design to be made.

They also show a wider range of student drawing ability and complement the ideation pages.

As mentioned earlier, before these drawings can be attempted, comprehensive classwork exercises must be undertaken, directed at every step by the teacher.

Third angle orthographic projection needs to be introduced in earlier years of study, so that students are not attempting to draw their designs in this format from the outset without any prior knowledge.

This drawing shows a wide range of essential drawing skills that include dimensioning, cutting planes, cross-hatching and a parts list.

Because of the complicated amount of detail in Anton's design, he has introduced a tracing film overlay to show sectioned information in the front elevation.

SKILLS NEEDED TO PRECEDE THIS PAGE:

Line weights and line standards (2H pencil for all lines), title block, scales and projection symbol.
Printing standards (HB pencil).
Correct dimensioning techniques for third angle projection.
Cutting plane and cross-hatching.
Hidden detail, reference lines, notation.
A parts list with numbered components about the views.
Views labelled.
Links to: pages 71, 184–185

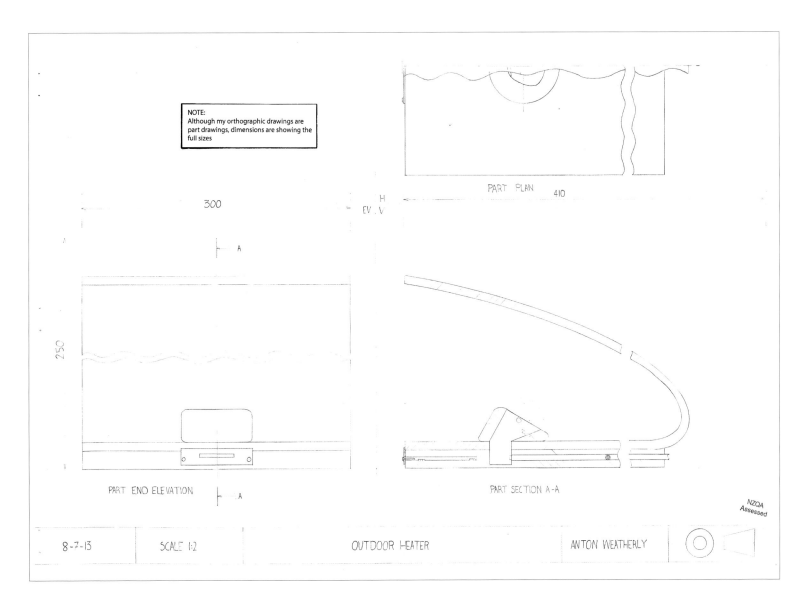

NOTE:
Although my orthographic drawings are
part drawings, dimensions are showing the
full sizes

PART PLAN   410

300

H
EV . V

A

250

PART END ELEVATION   A

PART SECTION A-A

NZQA
Assessed

8-7-13   SCALE 1:2   OUTDOOR HEATER   ANTON WEATHERLY

As I said on my first instrumental drawing page, these are not my main area of skill. However, I did try much harder to get the right hierarchy of lines on the page.

I practised this page three times to get the right scale and layout, but it still did not fit.

I decided, after talking to my teacher, to cut some of the front elevation out, and show the missing details on a tracing film overlay. This made the main drawing smaller and it then fitted the page. But it was a bit complicated showing the break symbols in the plan and front elevation.

I used a clutch pencil with a 2H lead for the lines and a 0.05 black fineline ink pen on the tracing film.

I took a long time about making sure I used the right layouts for writing the dimensions, the names of the views and for showing the cutting plane.

Even though the drawing took a long time, I am pleased with the way it turned out.

Nicole Dealey

The longest part about this drawing was working out the scale to fit the page. It took a few goes by practising on another page first, but the 1:20 scale fitted well in the end.

I took time over getting my lines right, starting by doing all of them in construction first, using a 2H pencil. I'm glad I had a clutch pencil for this.

I left out some of the hidden detail to make the drawing clearer, although I think my sectioned parts are a bit small to see well.

When I had outlined all the views, putting all the sizes and labels around the views took the next longest because I wanted to get them as accurate as I could. I used an HB lead in my clutch pencil for all the printing and in the title block. Using guide lines helped get the heights of the letters looking the same.

If I was to do it again I would only show one end elevation so that I could make the drawing larger and easier to see all the detail, especially the removed section.

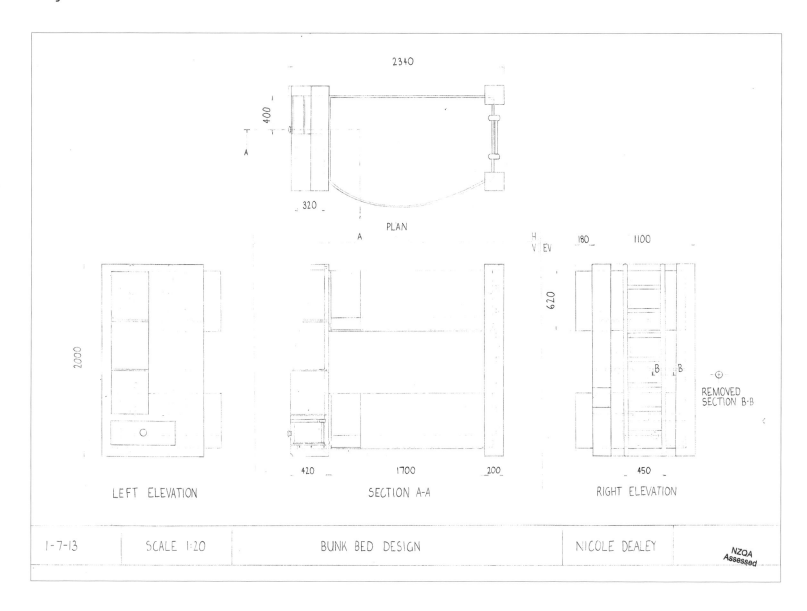

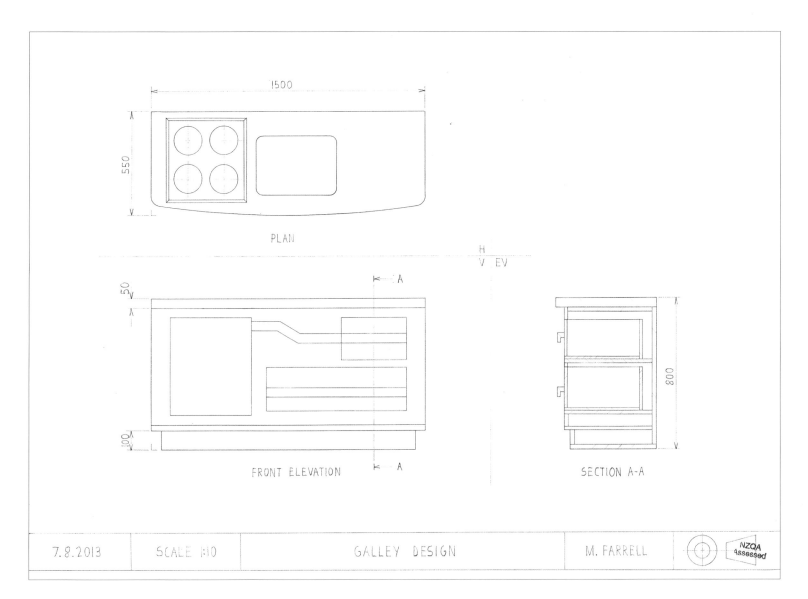

1500

550

PLAN

H
V EV

50

A

A

100

FRONT ELEVATION

A

A

800

SECTION A-A

7.8.2013	SCALE 1:10	GALLEY DESIGN	M. FARRELL	NZQA Assessed

After having already done an orthographic drawing for this brief, and because this was quite a simple design, it didn't take as long as I thought it would.

The skills I needed were the ones I knew about and had used in Year 10. I'm glad I had these because it made all the difference to the quality of the drawing.

Before I started I planned out how many views I needed and how the drawing would be positioned on the page.

After I had the drawing done in construction, I spent time checking that everything was correct before outlining. I took care to make sure I showed the right line hierarchy. I found doing medium-weight lines for the reference lines and cross-hatching the hardest.

I double-checked how to do the dimensions and cutting plane because in the past I always got them wrong.

I used a circle template instead of a compass for the four circles in the plan.

SITUATION

Traditionally and culturally, Kiwi summer holidays take place in the great outdoors. It's a time to experience the laid-back nature of New Zealand life and travel to hidden corners of our country.

BRIEF: Design a mobile home and a feature for the home.
DURATION: 20 weeks (class time)

REQUIREMENTS   **PART TWO: PRODUCT DESIGN**

1   Design research. *Cut and pasted among your sketches.*
2   Ideation of the feature. *Show: thumbnails, explorative sketches, thinking and technical sketches. Three A3 pages minimum.*
3   The feature must have a compound curve, circular parts and/or angled parts that can be plotted in instrumental drawings.
4   Place notes at any stage, about your research and sketches, to explain thinking and details that are not clear visually.
5   Demonstrate in your sketches the effective development of your design ideas, *e.g. explore alternative ideas, not limited to only the brief and its requirements, then refine your chosen ideas informed by aesthetics and function. Support design judgements with qualitative and/or quantitative data gained through research, reflecting your values, tastes and/or views. Where appropriate, use drawings, models, digital modelling, etc.*
6   Draw with instruments, a detailed, scaled third angle orthographic projection of your feature design. Show the following:
    - A minimum of three views, fully dimensioned.
    - Section one elevation to show construction detail.
    - A parts list with numbered components about the views.
    - Hidden detail, cross-hatching, notation and reference lines.
    - A title block with the scale and projection symbol.
    - Correct drawing standards and conventions.

7   Draw with instruments, a scaled paraline drawing of your design (oblique or isometric) and an exploded paraline drawing of a part of the design, *i.e. the joining method found in the construction of the part. Show a parts list and all constructions clearly.*

8   Using media of your choice, produce a hand-rendered presentation drawing of your design to show shape and surface qualities.

DESIGN SPECIFICATIONS

- The feature could be: *a bench/sink unit, bunks, table, outdoor seat, GPS, shelving, outdoor barbecue, etc.*
- The feature must: *consider human factors and be designed for either the interior or the exterior of the home.*
- It should consider both aesthetics and function.
  Function: construction, safety, user friendliness, fitness for purpose, ergonomics, etc.
  Aesthetics: style, proportion, finish, harmony, form, etc.

Instrumental drawing (cont.)

PARALINE DRAWING

Here's a 3D drawing of the design, drawn in isometric. Again, this drawing system offers important insight into the form of the design that could allow it to be made.

Another aspect of student drawing ability is also demonstrated, particularly in the construction of curves and/or circles.

Asking for curved and/or angled parts to be incorporated into the design (requirement 3 in the brief) gives students opportunities to gain higher achievement at assessment time.

As always, before this drawing can be attempted, comprehensive classwork exercises must be undertaken, directed at every step by the teacher.

Paraline drawing systems need to be introduced in earlier years of study, so that students are not attempting to draw their designs without any prior knowledge.

Note also the addition of an auxiliary view (in this case, a photocopy of the curved part from the orthographic projection), which allows for the plotting of the curved part in isometric.

This is an essential skill that needs to be demonstrated for assessment purposes.

SKILLS NEEDED TO PRECEDE THIS PAGE:

Line weights and line standards (2H pencil for all lines), title block and scales.
Printing standards (HB pencil).
3D circle and curve constructions.
A parts list with numbered components about the views.
Views labelled.
Links to: pages 71, 186–187

   ISBN: 9780170355575

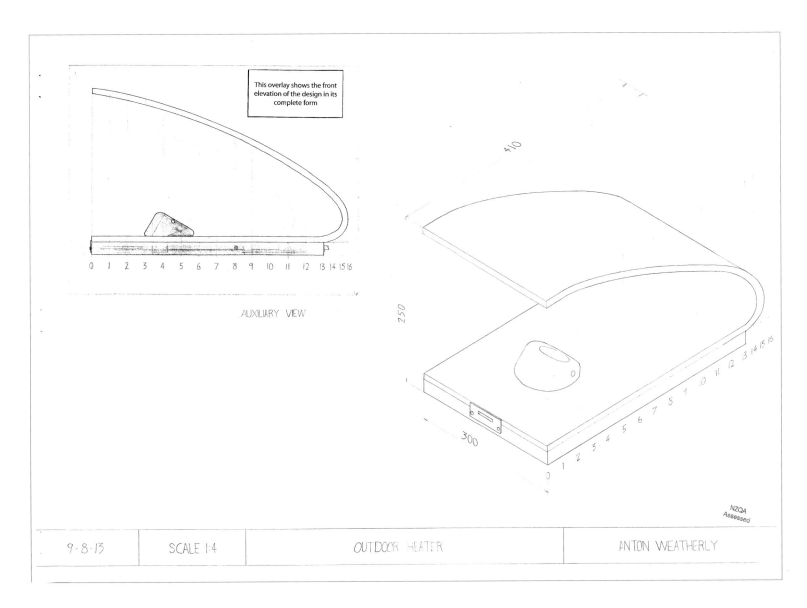

This overlay shows the front elevation of the design in its complete form

0 1 2 3 4 5 6 7 8 9 10 11 12 13 14 15 16

AUXILIARY VIEW

410

250

300

0 1 2 3 4 5 6 7 8 9 10 11 12 13 14 15 16

NZQA
Assessed

9-8-13	SCALE 1:4	OUTDOOR HEATER	ANTON WEATHERLY

This drawing took quite a bit of time to do, especially drawing the curve on the top in isometric using ordinates.

I feel, as with my orthographic projection drawings, that this drawing also lacks the proper linework. But I was a bit rushed for time so the difference between the construction lines and outlines are not as good as I would do if I did the page again.

I photocopied the front elevation of my orthographic projection and used this as the auxiliary view to draw the ordinates to plot the curve from.

It was good having learnt how to draw curves using ordinates in Year 10, so I knew how to do it.

If I was to do the drawing again, I would double-check all the sizes on the orthographic drawing to make sure they convert well to isometric.

Nicole Dealey

Although this drawing looks hard, and it took a long time to draw, I found it quite straightforward. I used the same scale as my orthographic projection so I could plot the curved side from it.

I started by photocopying the plan and front elevation of my orthographic projection. I used this as an auxiliary view to plot the curved sides of the bunks. I glued it to the page and put ordinates through the curve.

When the photocopy was glued on, I drew an isometric box, then put the shape inside it.

The longest part was using my compass to transfer the ordinates from the photocopy into the isometric drawing. I knew how to do this because I had done similar drawings in Year 10.

It took a long time to check the drawing before I outlined it, and a long time to outline it to get all the lines right. I used my clutch pencil with a 2H lead.

The dimensions were tricky to get them on the right angles.

If I was to do the drawing again, I would move it down to fit in the middle of the page.

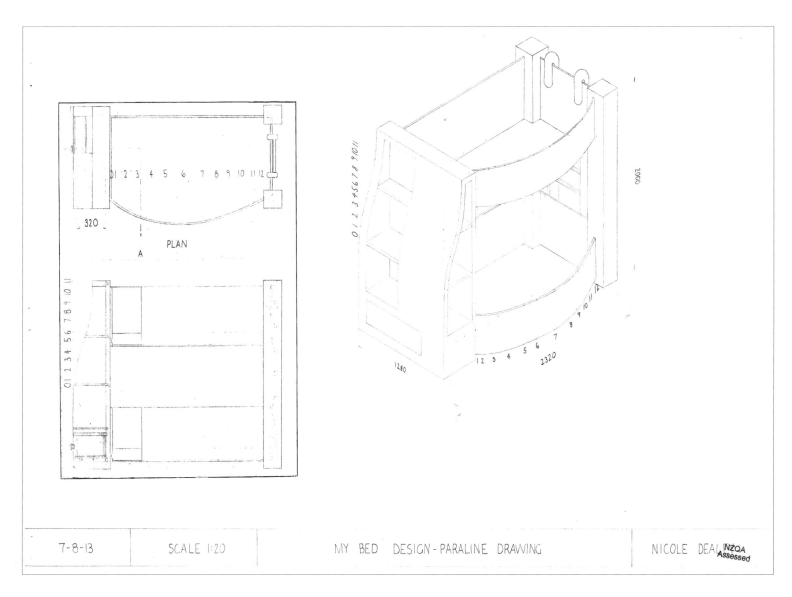

7-8-13	SCALE 1:20	MY BED DESIGN - PARALINE DRAWING	NICOLE DEAL

ISBN: 9780170355575

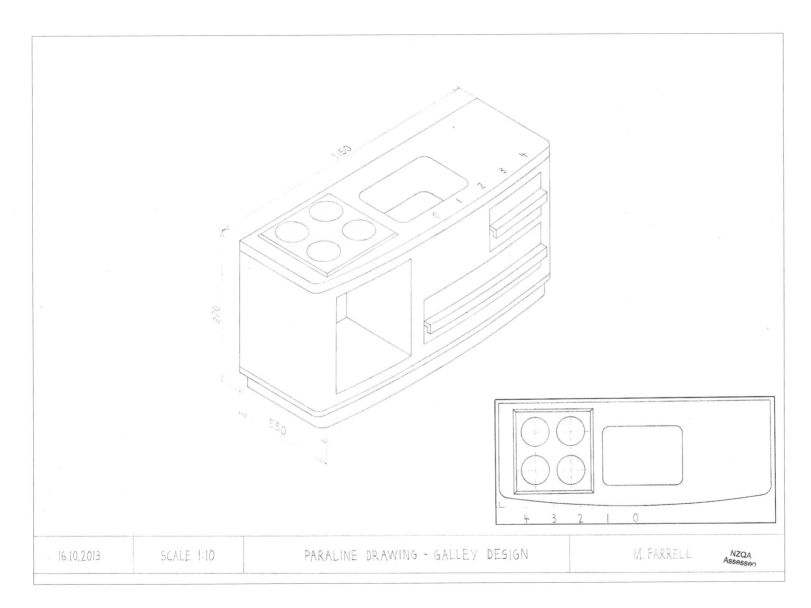

The first thing I did was make a photocopy of the plan of my orthographic projection. This meant that I didn't have to think about what scale to draw it in isometric, and it would make drawing the rounded front of the galley easier.

It did take a while to draw though, because of the ordinates and all the curved parts.

Apart from the curved front, which I plotted from the plan view, I used an ellipse template to draw most of the circles on the stove and the small rounded front corners.

I made sure that my lines were right, so I used a 2H lead in my clutch pencil.

The numbers and dimensions took me a while, getting them on the isometric angles. I did these last but used guide lines to make them neat.

I wanted to erase all the construction lines to make the drawing neater, but left them on to show how the drawing was done.

| 16.10.2013 | SCALE 1:10 | PARALINE DRAWING - GALLEY DESIGN | M. FARRELL | NZQA Assessed |

SITUATION

Traditionally and culturally, Kiwi summer holidays take place in the great outdoors. It's a time to experience the laid-back nature of New Zealand life and travel to hidden corners of our country.

BRIEF: Design a mobile home and a feature for the home.
DURATION: 20 weeks (class time)

REQUIREMENTS

PART TWO: PRODUCT DESIGN

1 Design research. *Cut and pasted among your sketches.*
2 Ideation of the feature. *Show: thumbnails, explorative sketches, thinking and technical sketches. Three A3 pages minimum.*
3 The feature must have a compound curve, circular parts and/or angled parts that can be plotted in instrumental drawings.
4 Place notes at any stage, about your research and sketches, to explain thinking and details that are not clear visually.
5 Demonstrate in your sketches the effective development of your design ideas, *e.g. explore alternative ideas, not limited to only the brief and its requirements, then refine your chosen ideas informed by aesthetics and function. Support design judgements with qualitative and/or quantitative data gained through research, reflecting your values, tastes and/or views. Where appropriate, use drawings, models, digital modelling, etc.*
6 Draw with instruments, a detailed, scaled third angle orthographic projection of your feature design. Show the following:
   ▪ A minimum of three views, fully dimensioned.
   ▪ Section one elevation to show construction detail.
   ▪ A parts list with numbered components about the views.
   ▪ Hidden detail, cross-hatching, notation and reference lines.
   ▪ A title block with the scale and projection symbol.
   ▪ Correct drawing standards and conventions.

7 Draw with instruments, a scaled paraline drawing of your design (oblique or isometric) and an exploded paraline drawing of a part of the design, *i.e. the joining method found in the construction of the part. Show a parts list and all constructions clearly.*

8 Using media of your choice, produce a hand-rendered presentation drawing of your design to show shape and surface qualities.

DESIGN SPECIFICATIONS

▪ The feature could be: *a bench/sink unit, bunks, table, outdoor seat, GPS, shelving, outdoor barbecue, etc.*
▪ The feature must: *consider human factors and be designed for either the interior or the exterior of the home.*
▪ It should consider both aesthetics and function.
   Function: construction, safety, user friendliness, fitness for purpose, ergonomics, etc.
   Aesthetics: style, proportion, finish, harmony, form, etc.

Instrumental drawing (cont.)

EXPLODED PARALINE DRAWING

This 3D drawing provides accurate, scaled information on the construction of main features of the design. Exploded isometric drawings confirm construction techniques that could allow the design to be manufactured. Four important factors need to be taken into account with these drawings:

▪ That the parts are exploded (pulled apart) in the direction in which they are assembled.
▪ That exploded lines and crating are clearly seen.
▪ That constructions of curved surfaces are clearly shown.
▪ That a parts list clearly identifies the parts.

As always, before these drawings can be attempted, comprehensive classwork exercises must be undertaken, directed at every step by the teacher.

Knowledge of exploded drawing techniques, and how to make a parts list, must be taught. Note that the parts list can be produced on a computer, then cut and pasted onto the drawing, as seen on the student work featured.

SKILLS NEEDED TO PRECEDE THIS PAGE:

Line weights and line standards (2H pencil for all lines).
Title block and scales.
Printing standards (HB pencil).
3D circle and curve constructions.
Exploded drawing techniques.
A parts list with numbered components about the views.
Links to: pages 71, 187–189

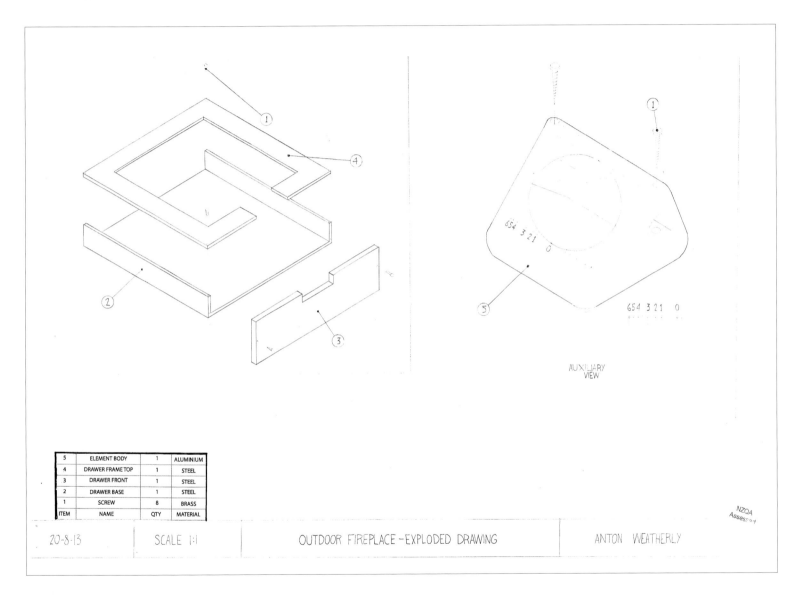

The hardest part about this drawing was deciding which parts to show exploded.

I did some practice drawings first and decided on the main parts and the ones easiest for me to draw. At the same time I had to make sure that the construction methods were correct, so I did a lot of research beforehand.

I feel that my linework is better in this drawing because I had done a lot of practice on the other instrumental drawings.

I made the parts list on a computer and pasted it onto the page. This kept the page neat and saved a lot of time instead of printing it by hand.

It took some time to work out scales to make sure the exploded parts would fit onto the page.

AUXILIARY VIEW

ITEM	NAME	QTY	MATERIAL
5	ELEMENT BODY	1	ALUMINIUM
4	DRAWER FRAME TOP	1	STEEL
3	DRAWER FRONT	1	STEEL
2	DRAWER BASE	1	STEEL
1	SCREW	8	BRASS

20-8-13   SCALE 1:1   OUTDOOR FIREPLACE – EXPLODED DRAWING   ANTON WEATHERLY

NZQA Assessed

Nicole Dealey

The hardest part about this drawing was working out which parts to show exploded. I did not have enough room on the page for the whole bunk bed, so I decided on drawing the main parts.

The next hardest part was finding out how the parts would be joined together. I researched this using the internet and books in my classroom.

Working out a scale for the drawings also took time. I did not want to make them too small and hard to see.

I practised doing some quick sketches before drawing them with instruments. Getting the drawings in the right place on the page was tricky. I did them in construction first but had to erase my first attempt.

I used my clutch pencil again with a 2H lead so that all my instrumental drawing pages looked at the same level.

The parts list was done on a computer and pasted on last.

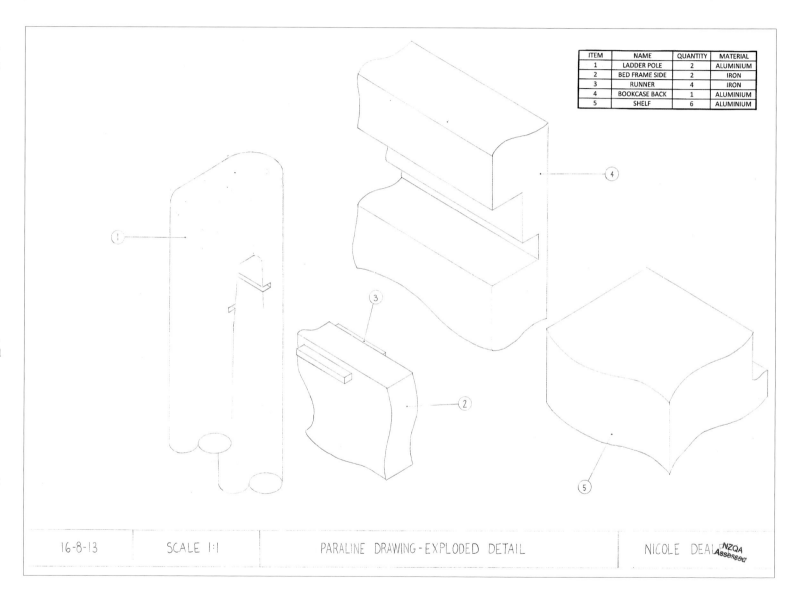

ITEM	NAME	QUANTITY	MATERIAL
1	LADDER POLE	2	ALUMINIUM
2	BED FRAME SIDE	2	IRON
3	RUNNER	4	IRON
4	BOOKCASE BACK	1	ALUMINIUM
5	SHELF	6	ALUMINIUM

16-8-13   SCALE 1:1   PARALINE DRAWING - EXPLODED DETAIL   NICOLE DEALEY

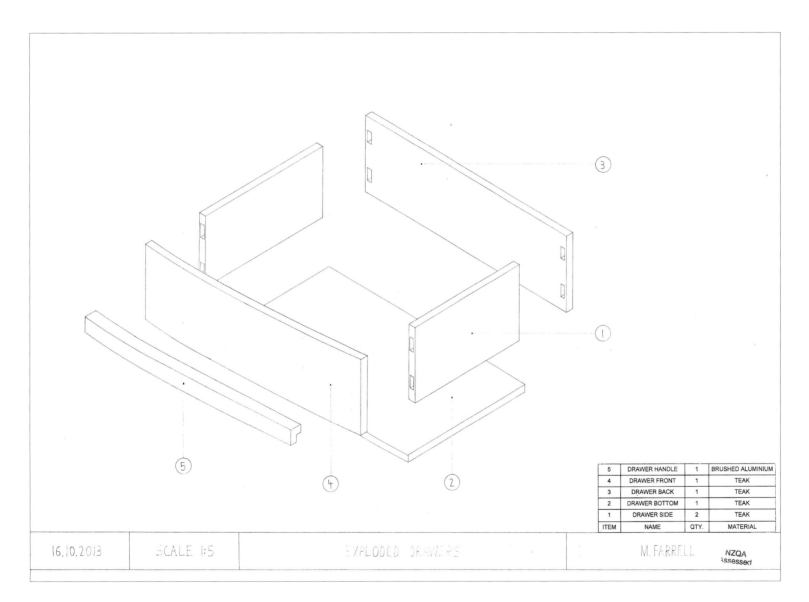

5	DRAWER HANDLE	1	BRUSHED ALUMINIUM
4	DRAWER FRONT	1	TEAK
3	DRAWER BACK	1	TEAK
2	DRAWER BOTTOM	1	TEAK
1	DRAWER SIDE	2	TEAK
ITEM	NAME	QTY.	MATERIAL

16.10.2013	SCALE 1:5	EXPLODED DRAWERS	M. FARRELL	NZQA Assessed

I needed to speed my pages up at this stage because I was running out of time. So I kept this drawing simple by only showing one exploded detail of the drawer.

I looked in some building books in my classroom and on the internet to find out how it could be made.

The hardest part was drawing the curved front of the drawer. I drew it inside a crate first, then used ordinates to plot it.

The scale was easy to work out because I only had one drawing on the page.

I took time to make sure that my linework was the same as my other drawings so that the quality matched. I used a 2H pencil for all the lines and an HB pencil for printing.

The parts list was made on a computer and pasted onto the title block at the bottom of the page.

SITUATION

Traditionally and culturally, Kiwi summer holidays take place in the great outdoors. It's a time to experience the laid-back nature of New Zealand life and travel to hidden corners of our country.

BRIEF: Design a mobile home and a feature for the home.
DURATION: 20 weeks (class time)

REQUIREMENTS   **PART TWO: PRODUCT DESIGN**

1   Design research. *Cut and pasted among your sketches.*
2   Ideation of the feature. *Show: thumbnails, explorative sketches, thinking and technical sketches. Three A3 pages minimum.*
3   The feature must have a compound curve, circular parts and/or angled parts that can be plotted in instrumental drawings.
4   Place notes at any stage, about your research and sketches, to explain thinking and details that are not clear visually.
5   Demonstrate in your sketches the effective development of your design ideas, *e.g. explore alternative ideas, not limited to only the brief and its requirements, then refine your chosen ideas informed by aesthetics and function. Support design judgements with qualitative and/or quantitative data gained through research, reflecting your values, tastes and/or views. Where appropriate, use drawings, models, digital modelling, etc.*
6   Draw with instruments, a detailed, scaled third angle orthographic projection of your feature design. Show the following:
    - A minimum of three views, fully dimensioned.
    - Section one elevation to show construction detail.
    - A parts list with numbered components about the views.
    - Hidden detail, cross-hatching, notation and reference lines.
    - A title block with the scale and projection symbol.
    - Correct drawing standards and conventions.
7   Draw with instruments, a scaled paraline drawing of your design (oblique or isometric) and an exploded paraline drawing of a part of the design, *i.e. the joining method found in the construction of the part. Show a parts list and all constructions clearly.*

8   Using media of your choice, produce a hand-rendered presentation drawing of your design to show shape and surface qualities.

DESIGN SPECIFICATIONS

- The feature could be: *a bench/sink unit, bunks, table, outdoor seat, GPS, shelving, outdoor barbecue, etc.*
- The feature must: *consider human factors and be designed for either the interior or the exterior of the home.*
- It should consider both aesthetics and function.
  Function: construction, safety, user friendliness, fitness for purpose, ergonomics, etc.
  Aesthetics: style, proportion, finish, harmony, form, etc.

Presentation drawing

To ensure the 'wow' factor for this final drawing, the most important consideration is viewpoint. Look once again at the design sketches. There may be one that already has the right look and feel. If so, re-sketch it freehand, but neater, and with cleaner linework. *Hint: By sketching the presentation drawing freehand, and being bold in the use of viewpoint and angle of lines, a more 'styley' look can be achieved, rather than a 'clunky' instrumental drawing.*

Make several photocopies on cartridge paper (not the thin photocopy paper) to allow for the use of marker pens. Photocopying will also enable the drawing to be enlarged or reduced to suit. Preserve one copy as the master, on which final renderings and cut and pasted areas can be placed. Use the others for practice. Once the master is fully rendered and outlined, it can be cut out ready to be mounted onto a background.

Anton has used an airbrush to achieve the look for the curved top surfaces of his design, and marker pen for the flat areas of the base. White fineline gel pens add edge highlights and gouache adds sparkle to the corners. The background has been produced by smudging chalk pastel pencil within a masked rectangle drawn on a toothed card. Edges have been sharpened with an eraser. Eraser stripes also add visual interest through the pastel. White gel pen has been used for the drop border.

Remember that these techniques must be taught — they are not a given. Precede this drawing with exercises to give practice in applying a range of media to different surface textures and form that will ultimately inform student decisions about their own designs. Sound class management is required, as teachers will be called upon to give much individual help.

Hint: Get students to practise writing a 'styley' signature. The addition of a signature, in the right place, adds a personal and professional touch.

SKILLS NEEDED TO PRECEDE THIS PAGE:

Perspective drawing techniques, viewpoint, backgrounds.
Crating (proportions) and line hierarchy.
Rendering techniques — light source, tonal changes, shape and surface texture.
Colour palette considered.
Media technique: marker pens, black fineline pens, pastel pencils.
Links to: pages 82–91

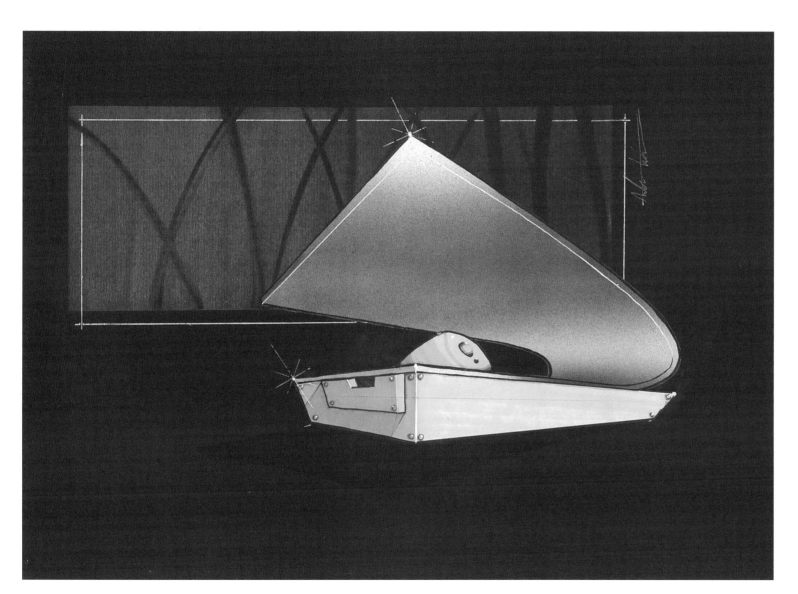

I used a viewpoint from one of my design sketches, as I wanted to show this in a more dramatic way.

I sketched the drawing first to get the viewpoint right, then made some photocopies.

On one photocopy I rendered the base part with grey markers, then airbrushed the top part on another photocopy.

Then I cut both parts out and glued them onto a new piece of paper, then cut the whole drawing out again.

I thought a lot about a background and practised several on a separate sheet. I wanted to represent the landscape in which the fire would be used, so settled on a stylised rectangle of smudged pastel with eraser stripes to represent trees.

I made the background on dark pastel paper, then spray-glued my fire on it. I used gouache to add sparkle to the corners and a white gel pen to show shine on the top corners. I finished the drawing with my signature.

Nicole Dealey

I wanted to make this drawing look a bit more stylish than the one I did for the mobile home.

I practised a lot of different drawing methods and spent time looking from different angles until I got it to look interesting.

I decided on using three-point perspective, and even though I was a bit worried that it might look overdone, I am pleased with the finished result.

I rendered the drawing using marker pens and a white pastel pencil smudged onto the rounded parts to make them look more curved.

White gel pen and gouache was used for the highlights on the edges and for the sparkle on the top corners.

I cut the drawing out and pasted it onto the background, which is pastel smudged onto a dark-blue pastel paper. To finish it I added a curve underneath with a white pastel pencil. My signature took longer than I thought it would to get it looking right.

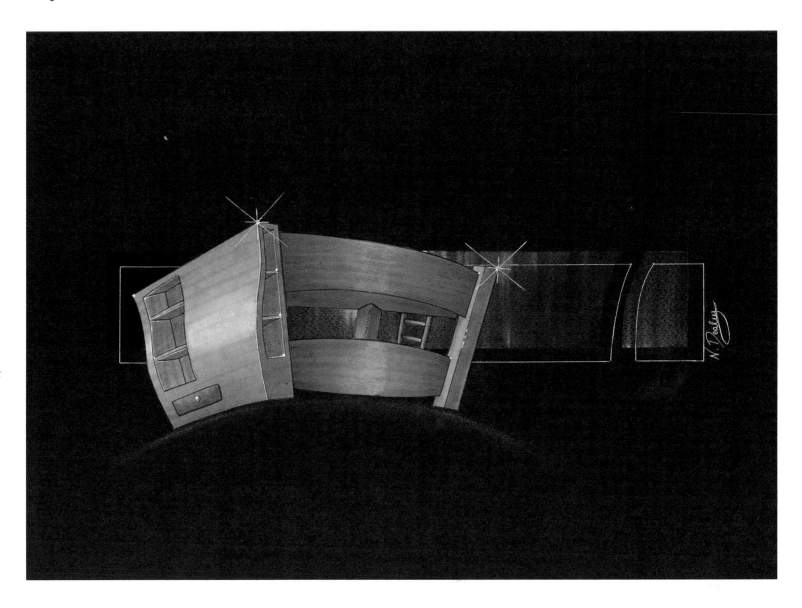

ISBN: 9780170355575

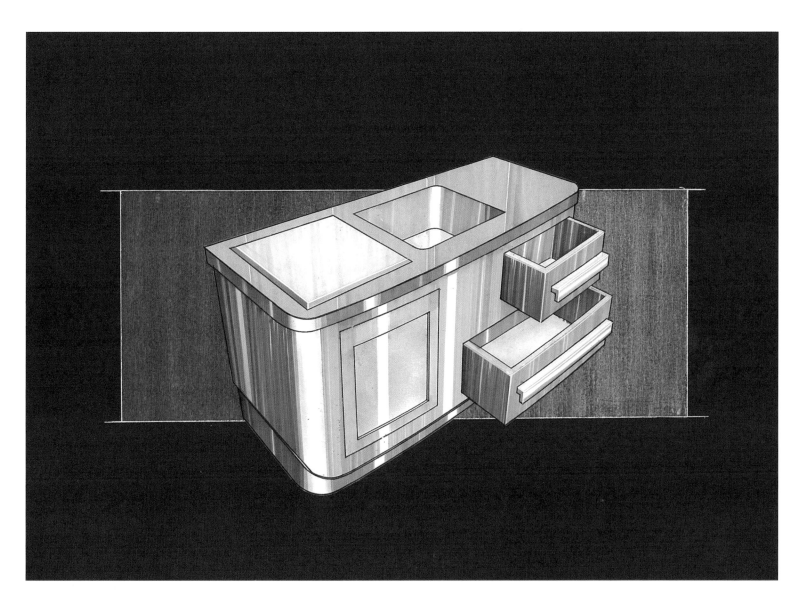

I found this drawing easier to do than the yacht, but it still took a long time.

I practised to get the right viewpoint and the right look for the different materials.

I used a three-point perspective view and although it looks okay, I would make the angles steeper if I was to do it again.

All the parts I rendered on separate sheets, then cut and pasted them together when I was happy with the look I wanted.

I used marker pens for the rendering. The colours were light greys and sandstone. The drawing was outlined with a black fineline pen.

I smudged white pastel onto textured paper for the background, using a frisket film mask to make the edges sharp, then pasted my drawing on top.

If I was to do it again I would make the background rectangle narrower and use gouache to add some sparkle.

BUILDING CONSTRUCTION

Framing details

The part pictorial view of a small wooden-framed building with a concrete foundation and floor, walls and roof are shown. The names of the main framing parts and their common sizes (in mm) are also shown.

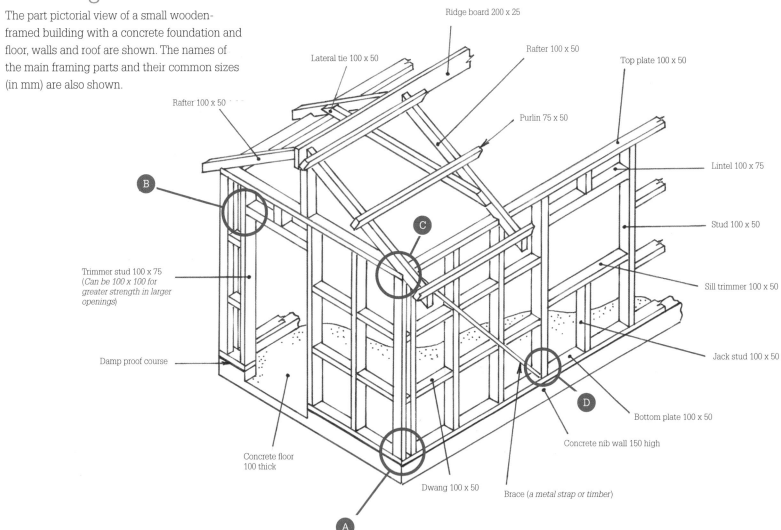

Ridge board 200 x 25

Rafter 100 x 50

Lateral tie 100 x 50

Top plate 100 x 50

Rafter 100 x 50

Purlin 75 x 50

Lintel 100 x 75

Stud 100 x 50

Trimmer stud 100 x 75 (*Can be 100 x 100 for greater strength in larger openings*)

Sill trimmer 100 x 50

Jack stud 100 x 50

Damp proof course

Bottom plate 100 x 50

Concrete nib wall 150 high

Concrete floor 100 thick

Dwang 100 x 50

Brace (*a metal strap or timber*)

ISBN: 9780170355575

DETAIL OF THE JOINTS USED BETWEEN FRAMING PARTS OF A BUILDING

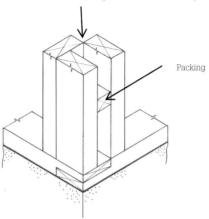

Corner allows for the fixing of the interior wall lining

Packing

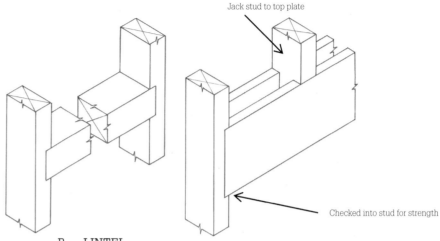

Jack stud to top plate

Checked into stud for strength

A — CORNER OF BUILDING

Three studs provide strength and form an inside corner for fixing the interior wall lining.

B — LINTEL

Used at the top of a window and door opening. A lintel can be solid (left) or built up (right) made from two thin boards to give more strength to a wide opening.

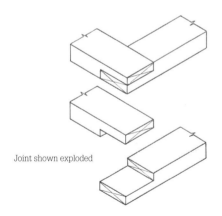

Joint shown exploded

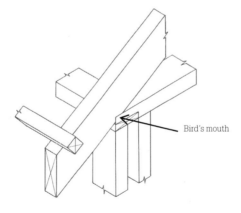

Bird's mouth

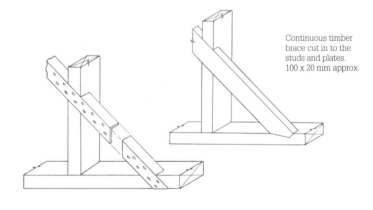

Continuous timber brace cut in to the studs and plates. 100 x 20 mm approx.

C — ANGLED HALVING JOINT

Equal parts of the top and bottom plates are cut at the ends to allow them to overlap and form a right-angled corner.

C — BIRD'S MOUTH JOINT

A cut is made in the rafter to allow it to rest on and be fixed securely to the top plate.

D — BRACE

Can be made from timber or metal. A metal angle brace (above left) is approximately 30 mm wide and is common in today's buildings. It is nailed into the studs and plates through holes provided down one side. The brace should be close to 45° for optimum strength.

Concrete foundation details

3D AND 2D SECTIONED VIEW OF TYPICAL CONCRETE FOUNDATION AND FLOOR

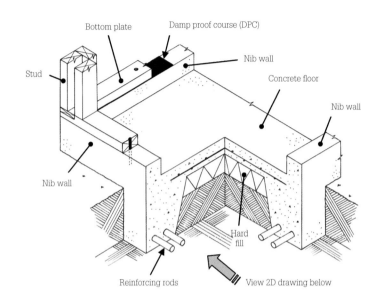

Bottom plate
Damp proof course (DPC)
Nib wall
Concrete floor
Nib wall
Stud
Nib wall
Hard fill
Reinforcing rods
View 2D drawing below

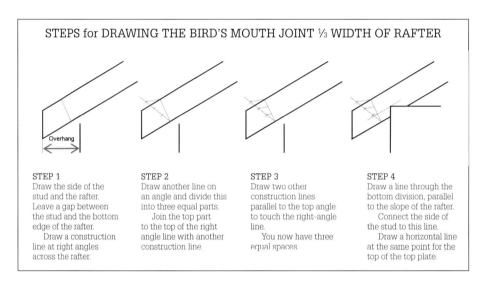

STEPS for DRAWING THE BIRD'S MOUTH JOINT ⅓ WIDTH OF RAFTER

Overhang

STEP 1
Draw the side of the stud and the rafter. Leave a gap between the stud and the bottom edge of the rafter.
Draw a construction line at right angles across the rafter.

STEP 2
Draw another line on an angle and divide this into three equal parts.
Join the top part to the top of the right angle line with another construction line.

STEP 3
Draw two other construction lines parallel to the top angle to touch the right-angle line.
You now have three equal spaces.

STEP 4
Draw a line through the bottom division, parallel to the slope of the rafter.
Connect the side of the stud to this line.
Draw a horizontal line at the same point for the top of the top plate.

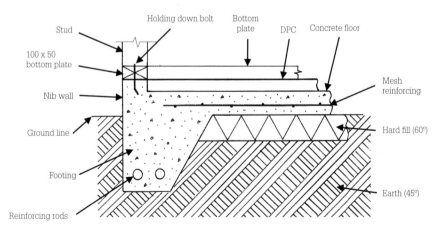

Stud
Holding down bolt
Bottom plate
DPC
Concrete floor
100 x 50 bottom plate
Nib wall
Ground line
Footing
Reinforcing rods
Mesh reinforcing
Hard fill (60°)
Earth (45°)

CROSS-SECTION VIEW THROUGH FOUNDATION

STEPS FOR DRAWING THE CROSS-SECTION

STEP 1
Draw the line for the corner of the building (stud).

STEP 2
Draw the ground line.

STEP 3
Where the ground line and corner of building meet, measure the height of the nib wall and the depth of the concrete footing. Step out the sizes of all other parts.

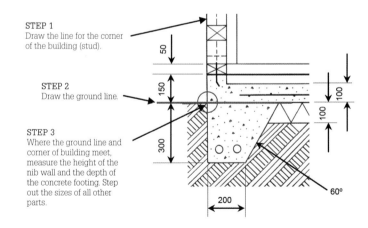

50
150
300
100
100
200
60°

ISBN: 9780170355575

Do these exercises to develop the following building drawing skills:

- Line types — construction, outlines, centre lines.
- The names of the parts of the timber framework of a building.
- The scale of the drawing and how to write it: 1:10, 1:20, 1.50, 1.100, etc.
- How to show dimensions (the sizes of the object).
- How to draw a title block, cutting plane and sectioned surfaces.
- How to draw material symbols in sectioned views and timber break symbols.

EXERCISE 1: CORNER OF BUILDING

The oblique drawing of the corner of the timber framework for a building is given.

1 Using instruments, to a scale of 1:10, draw the sectioned end elevation on cutting plane A-A. *Show the concrete wall and floor, hard fill, earth, the footing with two Ø12 reinforcing rods, and construction of the bird's mouth joint where the rafter rests on the top plate (rafter thickness).* See previous page. Follow the numbered steps 1–3 in the page layout below.

2 Use the sizes in the specifications chart.

3 Project the front elevation. Make the wall height to fit your page. *Show timber and concrete break symbols.*

Show:

- A title block and the projection symbol.
- The cutting plane and all cross-section symbols.
- Four dimensions.

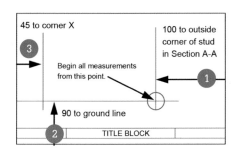

SPECIFICATIONS	
Framing	100 x 50
Nib wall	150 high
Footing	300 x 200
Floor	100 thick
Hard fill	100 thick
Brace (metal angle brace)	30 mm 45°
Rafters	100 x 50. 300 mm overhang. 30° pitch (angle)
Purlin	75 x 50

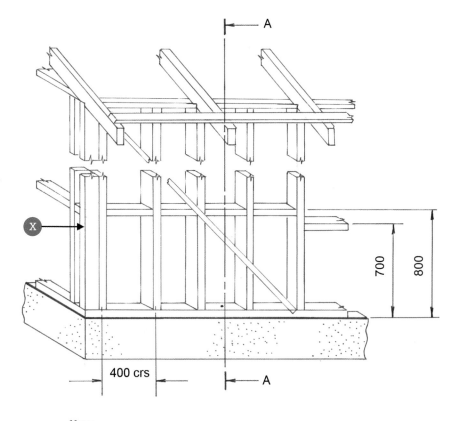

Note:
crs stands for centres.

See answers on page 212.

EXERCISE 2: ROOF FRAMING

The orthographic projection of roof framing parts are given.

Working from the page layout below, use instruments to redraw the framing in isometric when looking up and in the direction of the arrows. *For extension, you could show the lintel exploded where it joins the stud.*

Use a scale of 1:5.

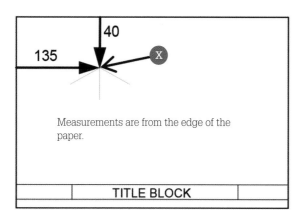

Measurements are from the edge of the paper.

TITLE BLOCK

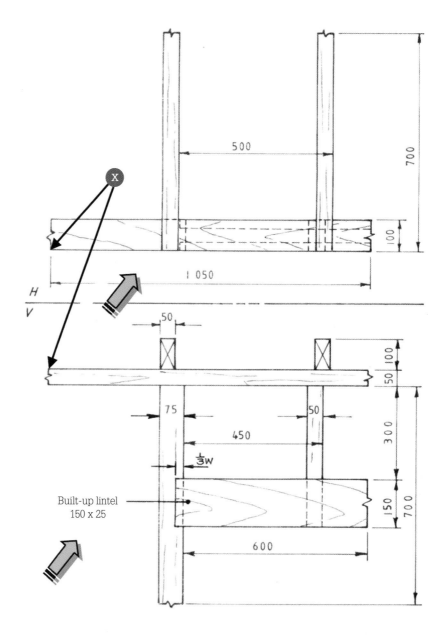

Built-up lintel
150 x 25

See answers on page 212.

Architectural symbols

Shown below are material symbols, as seen on cross-sections, and electrical wiring symbols. They should mostly be drawn with instruments and templates. Some parts may be drawn neatly freehand. Sizes should be appropriate to the size of the drawing.

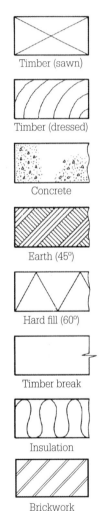

Timber (sawn)

Timber (dressed)

Concrete

Earth (45°)

Hard fill (60°)

Timber break

Insulation

Brickwork

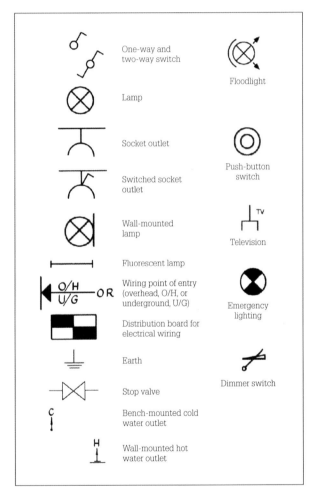

One-way and two-way switch

Lamp

Socket outlet

Switched socket outlet

Wall-mounted lamp

Fluorescent lamp

Wiring point of entry (overhead, O/H, or underground, U/G)

Distribution board for electrical wiring

Earth

Stop valve

Bench-mounted cold water outlet

Wall-mounted hot water outlet

Floodlight

Push-button switch

Television

Emergency lighting

Dimmer switch

COMMON SYMBOLS FOR WINDOWS, DOORS, PLUMBING FIXTURES, ETC.

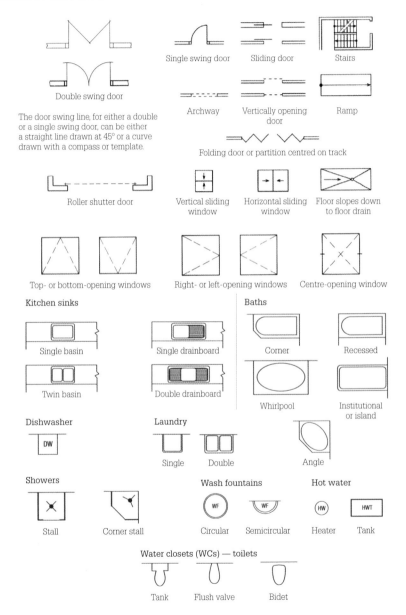

Single swing door

Sliding door

Stairs

Double swing door

Archway

Vertically opening door

Ramp

The door swing line, for either a double or a single swing door, can be either a straight line drawn at 45° or a curve drawn with a compass or template.

Folding door or partition centred on track

Roller shutter door

Vertical sliding window

Horizontal sliding window

Floor slopes down to floor drain

Top- or bottom-opening windows

Right- or left-opening windows

Centre-opening window

Kitchen sinks

Single basin

Single drainboard

Twin basin

Double drainboard

Baths

Corner

Recessed

Whirlpool

Institutional or island

Dishwasher

DW

Laundry

Single

Double

Angle

Showers

Stall

Corner stall

Wash fountains

WF

Circular

WF

Semicircular

Hot water

HW

Heater

HWT

Tank

Water closets (WCs) — toilets

Tank

Flush valve

Bidet

Do these exercises to develop the following orthographic projection skills:

- Setting out the drawing on the page and the correct line types to use.
- Making a title block with correct printing standards.
- Labelling the views, drawing reference lines and the projection symbol.
- Drawing to scale and how to write it: 1:1, 1:2, 1:5, 2:1, 1:10, etc.
- Showing dimensions about the views and metal break symbols.
- Constructing angles with a compass.
- Drawing curves and circles with a compass.
- Drawing cutting planes and cross-hatching on sectioned surfaces.

EXERCISE 1: STORAGE RACK

The isometric drawing (top right) and part end elevation (below) of a storage rack are shown.

1. To a scale of 1:5, redraw the storage rack in third angle orthographic projection. Show:
   - A plan and front elevation when viewed in the direction of the arrow. *Do not show hidden detail.*
   - A sectioned end elevation on cutting plane B-B. *The cutting plane passes through the centre rods.*
   - Six main dimensions, the reference line, notation and the views labelled.

2. Use a compass to construct the 105° angles of the sides and a recognised method to obtain the thickness parallel to the 105° angles. Show constructions clearly — see next page.

3. Draw the small elliptical curves of the rod ends with an ellipse template.

4. Leave projection lines clearly visible. *Hint: Begin by drawing the end elevation first.*

5. Draw and render two freehand, 3D sketches of a new design for the sides of the rack.

See answers on page 213.

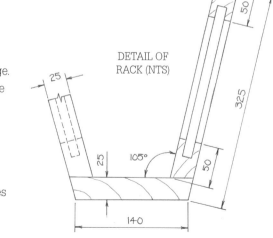

DETAIL OF RACK (NTS)

25

50

325

25

105°

50

140

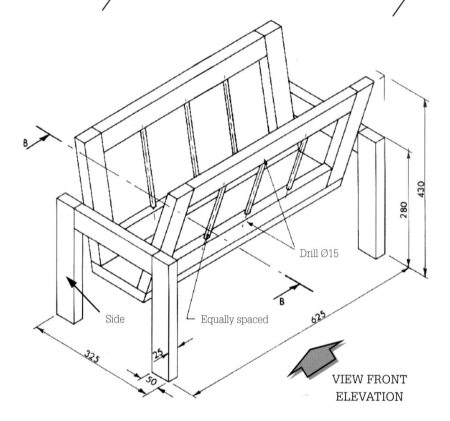

B

Drill Ø15

Side

Equally spaced

430

280

625

325

25

50

VIEW FRONT
ELEVATION

SPECIFICATIONS	
Frame members	50 x 25
Rods	250 mm long

MATERIALS	
Frame members	Radiata pine
Rods	Aluminium

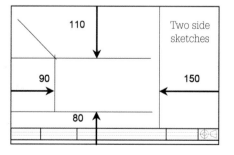

110

90

80

150

Two side sketches

ISBN: 9780170355575   PHOTOCOPYING OF THIS PAGE IS RESTRICTED UNDER LAW.

CONSTRUCTION TECHNIQUE FOR END OF STORAGE RACK

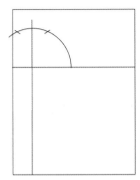

STEP 1
Draw the rectangle to contain the end elevation. Locate the point where the sloping rack begins and draw a semicircle. Mark the radius (60°) around the semicircle where it touches the top of the side.

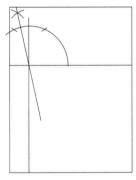

STEP 2
Bisect the 30° angle to obtain 15°. Draw the side of the frame which is now at 105°.

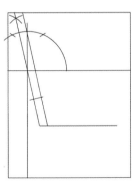

STEP 3
Draw a line at 90° to the 105° angle. Measure the thickness of the end of the sloping rack along it. Draw the thickness with a line parallel to the first line.

EXERCISE 2: CRANK ARM

The isometric drawing of a crank arm made from cast steel is shown. The crank arm is to be bolted to a piece of solid steel, which is not shown.

1   To a scale of 1:5, redraw the arm in third angle orthographic projection. Show:

■ A plan, the sectioned front elevation on cutting plane A-A, and the sectioned end elevation on cutting plane B-B.

■ A short length of 100 mm thick x 300 mm wide steel, on which the crank arm is bolted. *Show a metal break symbol. Do not show the bolt.*

■ A short length of solid steel shaft projecting through each side of the R100 hole. *Show a break symbol on each end of the shaft.*

■ The orthographic projection symbol (in the title block), reference line, notation and all constructions.

■ Hidden detail.

■ Six dimensions, the views labelled, both cutting planes and cross-hatching.

2   Leave projection lines clearly visible.

Hint: Draw the centre lines first (see page setout), then construct the drawing about them.

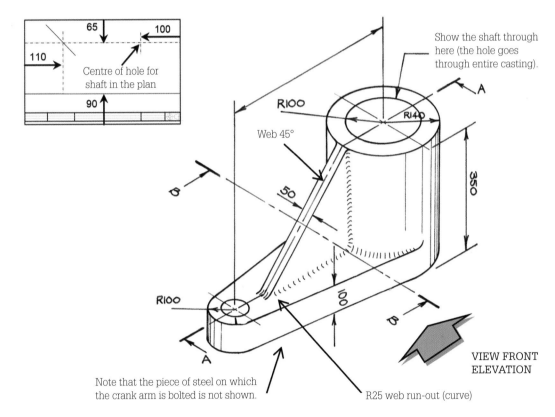

Show the shaft through here (the hole goes through entire casting).

Centre of hole for shaft in the plan

Web 45°

R100

R140

R100

R25 web run-out (curve)

VIEW FRONT ELEVATION

Note that the piece of steel on which the crank arm is bolted is not shown.

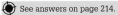 See answers on page 214.

EXERCISES

Do these exercises to develop the following isometric drawing skills:

- Setting out the drawing on the page and the correct line types to use.
- Making a title block with correct printing standards.
- Drawing to scale and how to write it: 1:1, 1:2, 1:5, 2:1, 1:10, etc.
- Converting from 2D to 3D and drawing from a given viewpoint.
- Showing dimensions about the views.
- Drawing curves and circles with a compass.

EXERCISE 1: PARKING METER

The orthographic projection of a parking meter is shown.

1 Use a scale of 1:2 to redraw the parking meter in isometric when looking up from beneath and in the direction of the arrows.

2 Show four main dimensions about the drawing and all constructions.

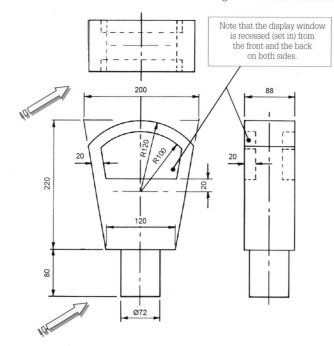

EXERCISE 2: LIGHT FITTING

The plan and front elevation of a light fitting and shade is shown.

1 Use a scale of 1:2 to redraw the light fitting in isometric viewed in the direction of the arrows.

2 Show four dimensions about the drawing and all constructions.

Note: An auxiliary view of the hexagonal shade (the plan) is required from which to plot the isometric drawing.

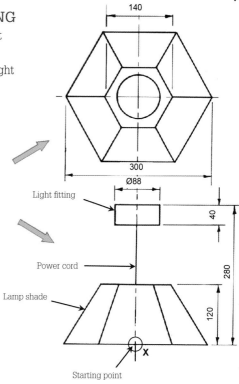

PARKING METER

Draw the centre lines through the middle of the parking meter first. Then construct the top curves.

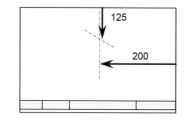

PAGE LAYOUTS

LIGHT FITTING

Starting point is the centre of bottom of lamp shade, X.

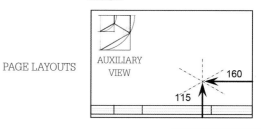

See answers on page 215.

ISBN: 9780170355575   PHOTOCOPYING OF THIS PAGE IS RESTRICTED UNDER LAW.

EXERCISE 3: WALL LAMP

The plan and front elevation of a wall lamp and bracket is shown.

1 On the left side of an A3 sheet, to a scale of 1:1, redraw the views as shown. (Use French curves to draw the compound curve of the bracket similar to that shown.)

2 On the right side of the page, draw an isometric view of the lamp and bracket when looking up and in the direction of the arrows.

3 Show all constructions clearly.

Notes

- You will need to place ordinates through the bracket curve in the front elevation.
- A number system is a good idea (see answer).
- Use dividers to transfer these to the isometric drawing.

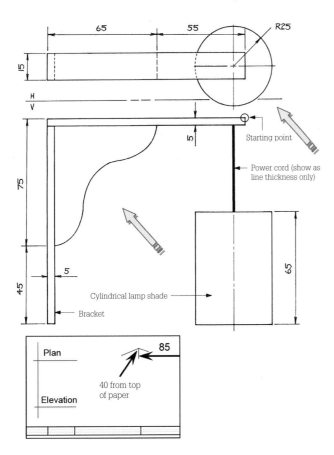

EXERCISE 4: EXPLODED DESK

The isometric drawing of a simple wooden desk is shown.

1 Study it carefully, and its sizes in the specifications chart.

2 On an A3 sheet, draw with instruments, to a scale of 1:10, isometric exploded views of the joints between parts 1 and 2, 2 and 3, 7, 8 and 9, and the drawer parts 4, 5, 6.

3 Draw a parts list (shown below) with numbered components about the views.

4 Show all constructions and exploded lines clearly.

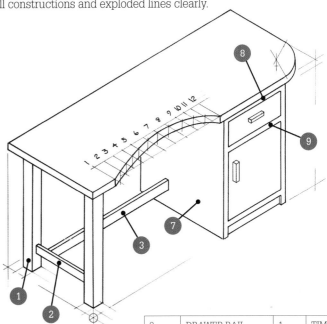

SPECIFICATIONS	
ALL TIMBER	20 mm thick
LEGS	70 mm square
LEG RAILS	60 mm x 20 mm
TOP RAIL AND DRAWER RAIL	40 mm x 20 mm
DRAWER SIDE AND BACK	120 mm x 6 mm thick
DRAWER BOTTOM	4 mm thick

ITEM	NAME	QTY	MATERIAL
9	DRAWER RAIL	1	TIMBER
8	TOP RAIL	1	TIMBER
7	CARCASE SIDE	2	TIMBER
6	DRAWER BOTTOM	1	MDF
5	DRAWER BACK	1	TIMBER
4	DRAWER SIDE	2	TIMBER
3	SPAN RAIL	1	TIMBER
2	RAIL	1	TIMBER
1	LEG	2	TIMBER

See answers on page 216.

Do these exercises to develop the following oblique drawing skills:

- Setting out the drawing on the page and the correct line types to use.
- Making a title block with correct printing standards.
- Drawing to scale and how to write it: 1:1, 1:2, 1:5, 2:1, 1:10, etc.
- Converting from a 2D to a 3D drawing.
- Showing dimensions about the views.
- Drawing curves, circles and angles.

EXERCISE 1: PENCIL SHARPENER

The orthographic projection of a pencil sharpener is shown.

1 Use a scale of 1:1 to redraw the sharpener in cavalier oblique when the end elevation is true shape. *Judge any sizes not given.*

2 Show four dimensions and all constructions.

3 Render the sharpener to show form *(the shape and surface qualities of the sharpener).*

4 Cut and paste the finished drawing onto a suitable background.

Hint: Begin by drawing the end elevation first, about the R30 centre line, then Part A and the handle about the R20 centre line.

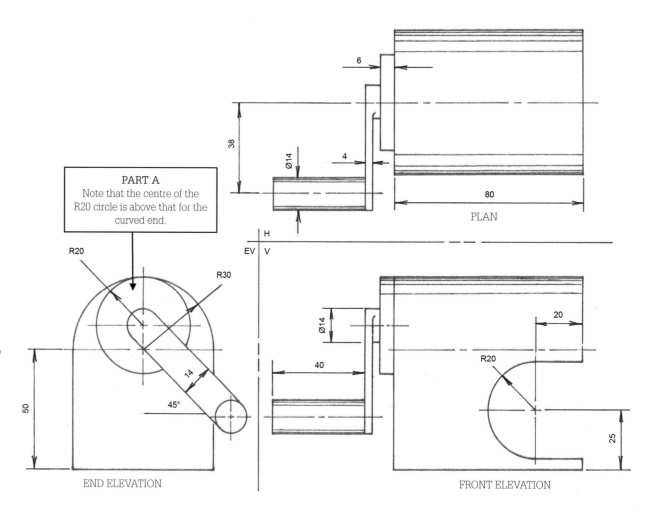

PART A
Note that the centre of the R20 circle is above that for the curved end.

END ELEVATION

PLAN

FRONT ELEVATION

See answers on page 217.

ISBN: 9780170355575   PHOTOCOPYING OF THIS PAGE IS RESTRICTED UNDER LAW.

EXERCISE 2: BIKE LAMP

The orthographic projection of a bike lamp is given. The sketch shows where it will be fitted to the handle bars.

1 Using a scale of 1:1 redraw the lamp in cavalier oblique when Face B is true shape, viewed in the direction of the arrow.

2 Show five dimensions and all constructions. *Judge any sizes not given.*

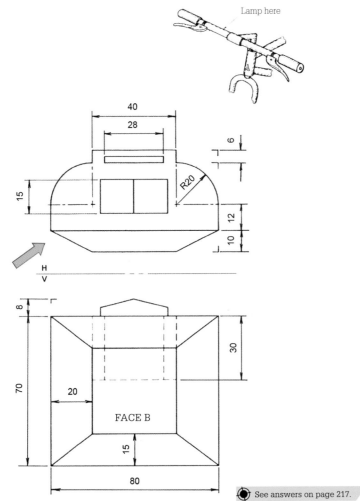

Do these exercises to develop the following perspective drawing skills:

- Line types — construction (visual rays), outlines.
- How to divide spaces equally, how to draw a shadow.
- How to draw in one-point (parallel) perspective and two-point (angular) perspective.
- How to draw the horizon line and vanishing points.
- How to draw landscape features and a human figure.
- How to render a dwelling and the surrounding landscape.

EXERCISE 1: BARN AND LANDSCAPE

1 Follow the numbered steps to carefully redraw the layout shown below, onto an A3 sheet. *Sizes are from the paper edge.*

2 Working from the given sizes, and looking closely at the construction drawing on the next page, complete the barn. *Judge any sizes not given.*

3 Draw the landscape to include four or five equally spaced trees and a shadow formed by the barn.

4 Show all constructions clearly and render the finished drawing.

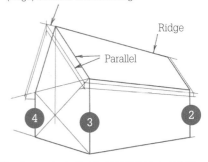

Extend the top of the roof (ridge) to make the overhang.

Ridge

Parallel

Numbers refer to the lines in the steps below.

See the next page for the layout and examples of the finished rendered drawing.

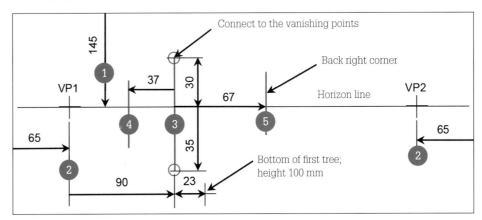

See answers on page 217.

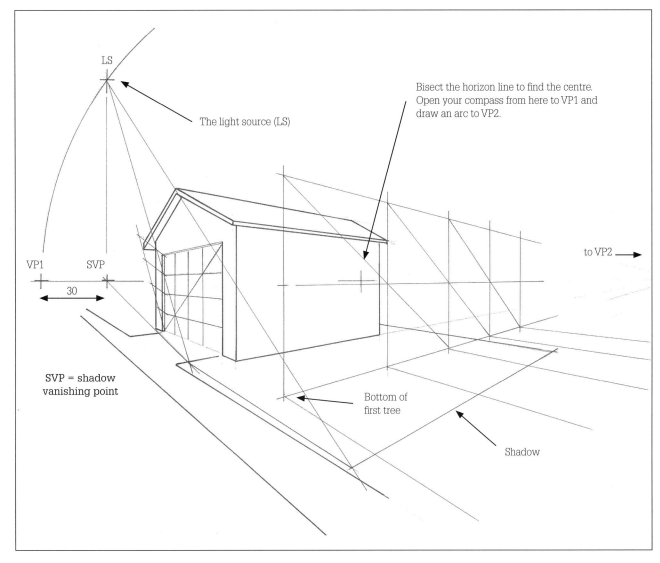

LS

The light source (LS)

Bisect the horizon line to find the centre.
Open your compass from here to VP1 and
draw an arc to VP2.

to VP2 →

VP1   SVP

30

SVP = shadow
vanishing point

Bottom of
first tree

Shadow

The drawings above show the use of
different rendering media for the barn and
landscape.
Top: colouring pencil and markers;
below: watercolour paint and markers.

Look carefully at the drawing above and the constructions required.

EXERCISE 2: HOUSE AND LANDSCAPE

1 Follow the numbered steps to carefully redraw the house shown below, onto an A3 sheet.
2 Use the details and features from the example given as a guide. Add any others you think appropriate.
3 Design, draw and render a landscape.

Remember: When setting out the windows, as they approach the vanishing points, they will become narrower and closer together.

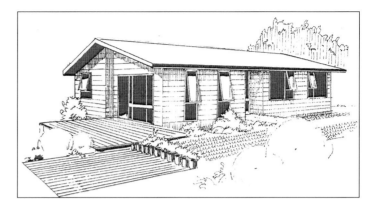

STEP 1
Draw the horizon line 135 mm from the top edge of the paper.

135

VP1

7

Horizon line

to VP2

16

20

50

STEP 2
Draw the VPs 7 mm from each side of the paper.

75

64

130

STEP 3

STEP 4

STEP 5

STEP 6
Draw visual rays to the VPs.

 See answers on page 218.

FREEHAND DESIGN SKETCHES (IDEATION)

Achieved

These design sketching examples (ideation) show the type of evidence required to achieve the standard. They should be read *as a guide only*.

Design sketching should show the following:

- Varied viewpoints
- Crating and proportion
- 2D and 3D sketches
- Alternative research
- Shadows
- Quick rendering from an identified light source to indicate materials, shape, form and finish (*aesthetics*)
- Line hierarchy (*thick and thin lines*)
- Details of construction (*exploded, sectional assembly, etc.*), function and/or movement
- Human factors
- Brief notes to discuss any detail that is not clear visually

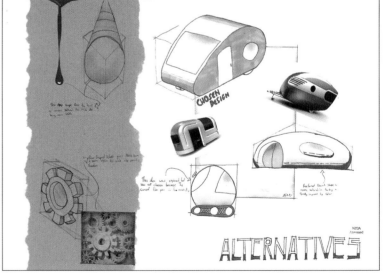

CHECKLIST	COMMENT
Viewpoints	The student has attempted to vary the viewing position of each sketch.
Crating and proportion	Crating is evident although many of the sketches lack correct proportion.
2D and 3D sketches	An attempt to show both sketch types with varying success. Some pages are too light on content and should show a wider range of ideas.
Alternative research	Some alternative research is beginning to inform the chosen design.
Quick rendering	Not really at a level that clearly indicates material. The student appears to have struggled when attempting to show form and finish. An identified light source has been taken into account.
Shadows	Not evident on these pages.
Line hierarchy	The student is aware of these and their importance to giving the sketch more visual appeal. A fair attempt has been made.
Construction detail	Not enough detail in the form of 3D exploded and 2D sectioned sketches. Arrows that show movement and/or function are not evident.
Human factors	A simple attempt at anthropometric data has been made.
Notes	Good notes with a focus on aesthetics and/or function.

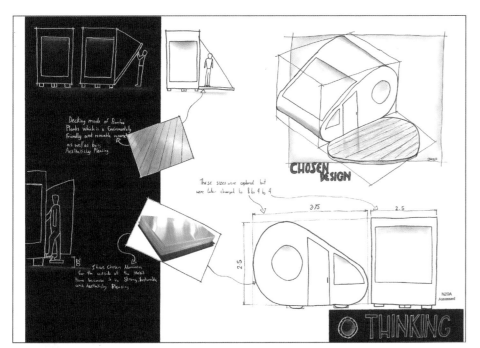

Decking made of Bamboo Planks which is a Environmentally friendly and renewable resource as well as being Aesthetically Pleasing

These sizes were explored but were later changed to 8 by 4 by 4

I have Chosen Aluminium for the outside of the Motel Home because it is Strong, Sustainable and Aesthetically Pleasing

CHOSEN DESIGN

3.75   2.5

2.5

THINKING

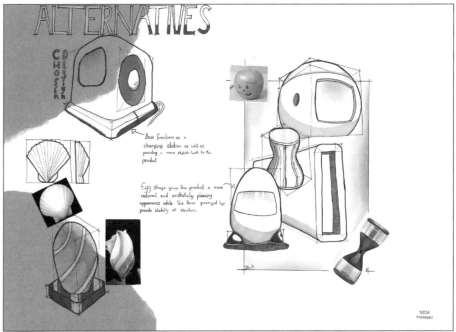

ALTERNATIVES

CHOSEN DESIGN

Base functions as a charging station as well as providing a more stylish look to the product

Egg shape gives the product a more natural and aesthetically pleasing appearance while the three pronged legs provide stability of structure.

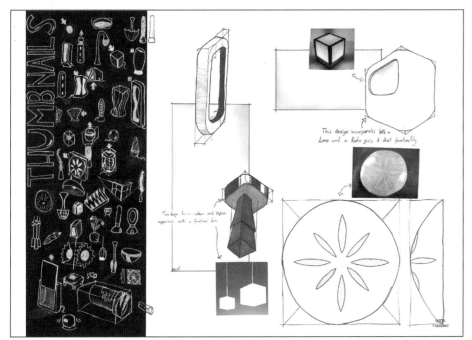

THUMBNAILS

This design incorporates both a Lamp and a Radio giving it dual functionality

This design has a modern and stylish appearance with a functional form

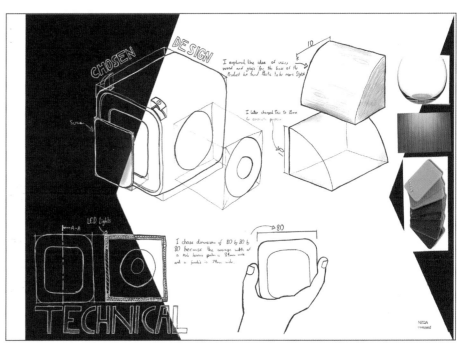

CHOSEN DESIGN

I explored the idea of using wood and glass for the base of the product but found Plastic to be more Stylish

I later changed this to 18mm for economic purpose

Screen

LED Lights

A–A

I chose dimensions of 80 by 80 by 80 because the average width of a Male human palm is 84mm wide and a female is 74mm wide.

80

TECHNICAL

Achieved with merit

These design sketching examples (ideation) show the type of evidence required to achieve the standard at a MERIT level. They should be read *as a guide only*.

Design sketching should show the following:

- Varied viewpoints
- Crating and proportion
- 2D and 3D sketches
- Alternative research
- Quick rendering from an identified light source to indicate materials, shape, form and finish (*aesthetics*)
- Shadows
- Line hierarchy (*thick and thin lines*)
- Details of construction (*exploded, sectional assembly, etc.*), function and/or movement
- Human factors
- Brief notes to discuss any detail that is not clear visually

CHECKLIST	COMMENT
Viewpoints	A range of different viewpoints gives a clear indication of design ideas.
Crating and proportion	Crating is clearly evident with well-proportioned design ideas.
2D and 3D sketches	The student has a good mix of 2D and 3D sketches that give a clear indication of design thinking.
Alternative research	Good use of alternative research in the form of organic shapes that better inform the chosen design.
Quick rendering	The student has used simple rendering to good effect to clarify materials, form and finish. An identified light source has been taken into account.
Shadows	Excellent use of shadows provides visual interest and 'anchors' the sketch to the page.
Line hierarchy	Excellent use of thick and thin lines makes the sketches 'pop' on the page.
Construction detail	A fine attempt at showing construction detail in the form of 3D exploded and 2D sectioned sketches.
Human factors	Good use of anthropometric data to achieve sizes.
Notes	Notes where necessary support the sketches.

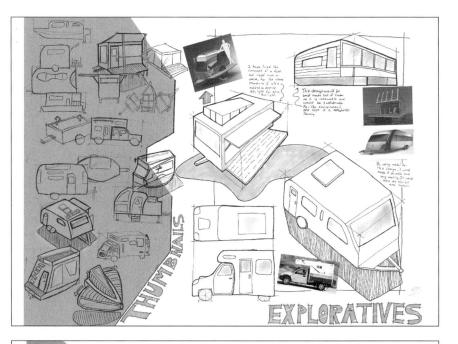

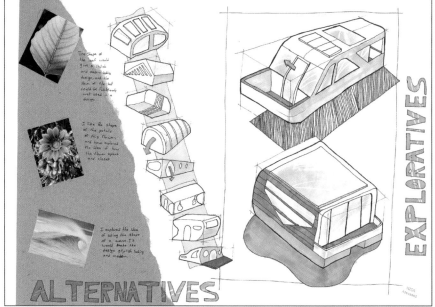

ISBN: 9780170355575

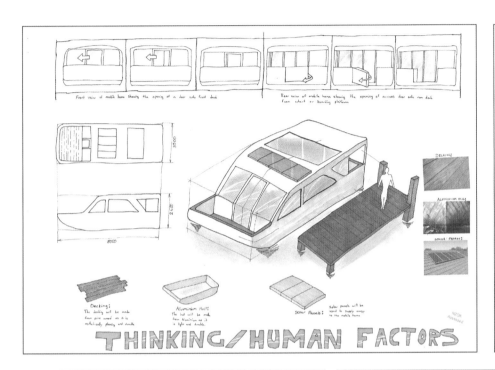

THINKING/HUMAN FACTORS

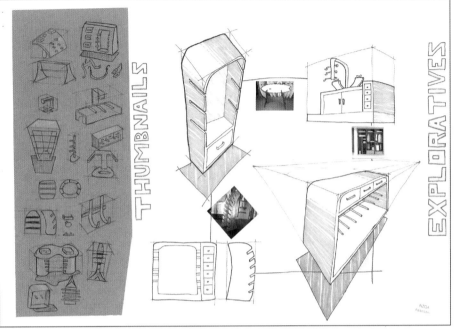

THUMBNAILS

EXPLORATIVES

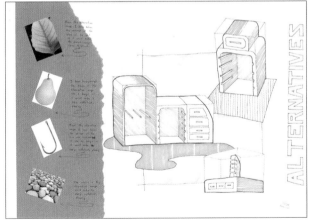

ALTERNATIVES

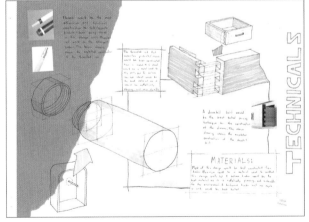

TECHNICALS

MATERIALS:

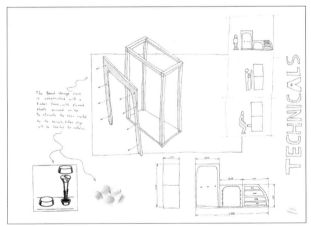

TECHNICALS

Achieved with excellence

These design sketching examples (ideation) show the type of evidence required to achieve the standard at an EXCELLENCE level. They should be read *as a guide only.*

Design sketching should show the following:

- Varied viewpoints
- Crating and proportion
- 2D and 3D sketches
- Alternative research
- Quick rendering from an identified light source to indicate materials, shape, form and finish (*aesthetics*)
- Shadows
- Line hierarchy (*thick and thin lines*)
- Details of construction (*exploded, sectional assembly, etc.*), function and/or movement
- Human factors
- Brief notes to discuss any detail that is not clear visually

CHECKLIST	COMMENT
Viewpoints	A wide range of different viewpoints gives an excellent and informative indication of design ideas.
Crating and proportion	Crating is clearly evident with well-proportioned design ideas.
2D and 3D sketches	A wide range of 2D and 3D sketches gives a clear indication of design thinking.
Alternative research	Superb use of alternative research clearly informs the chosen design.
Quick rendering	The student has used minimalist yet sophisticated rendering to clarify materials, form and finish. An identified light source has clearly been identified.
Shadows	Excellent use of shadows provides visual interest and 'anchors' the sketches to the page.
Line hierarchy	Excellent use of thick and thin lines makes the sketches 'pop' on the page.
Construction detail	Superb construction detail has been shown in the form of 3D exploded and 2D sectioned sketches.
Human factors	Excellent use of anthropometric data to achieve sizes.
Notes	Notes where necessary support the sketches.

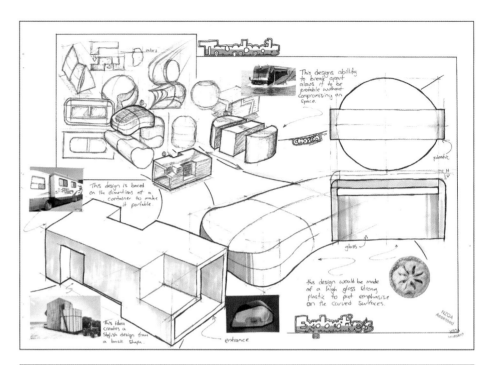

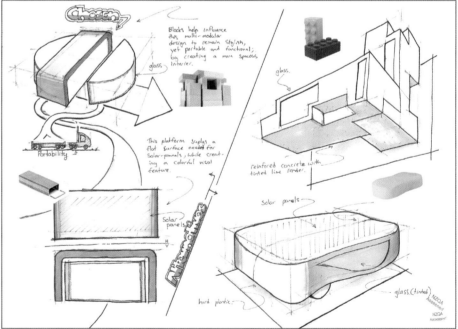

ISBN: 9780170355575   PHOTOCOPYING OF THIS PAGE IS RESTRICTED UNDER LAW.

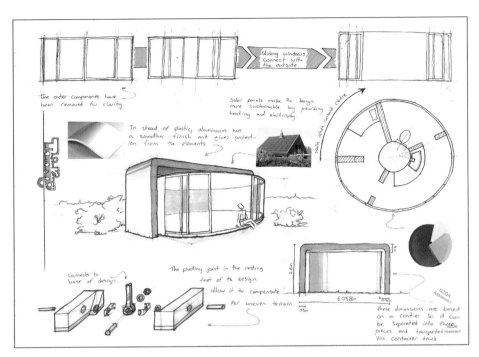

the order components have been removed for clarity.

Instead of plastic, aluminium has a smoother finish and gives protection from the elements.

Solar panels make the design more sustainable by providing heating and electricity.

Sliding windows connect with the outside.

connects to base of design.

The pivoting joint in the resting feet of the design allow it to compensate for uneven terrain.

6.058m

these dimensions are based on a cantier so it can be separated into three pieces and transported via container truck

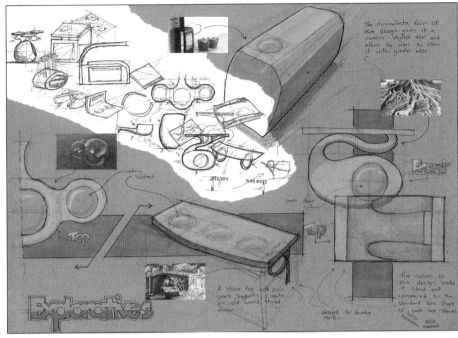

The minimalistic form of this design gives it a modern stylish feel and allows the user to clean it with greater ease.

Explorations

A stone top with iron work supports create an old world styled design.

The curves in this design make it stand out compared to the standard look shape of commercial top stoves.

designs to develop more...

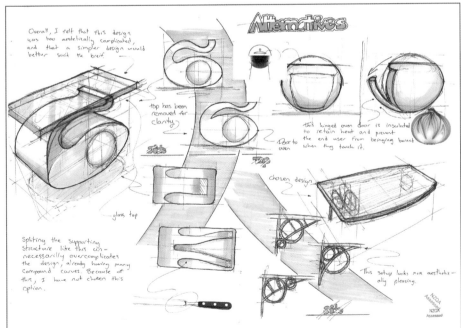

Alternatives

Overall, I felt that this design was too aesthetically complicated, and that a simpler design would better suit the brief.

top has been removed for clarity

Splitting the supporting structure like this unnecessarily overcomplicates the design, already having many compound curves. Because of this, I have not chosen this option.

glass top

this hinged oven door is insulated to retain heat and prevent the end user from being burned when they touch it.

chosen design

This setup looks more aesthetically pleasing.

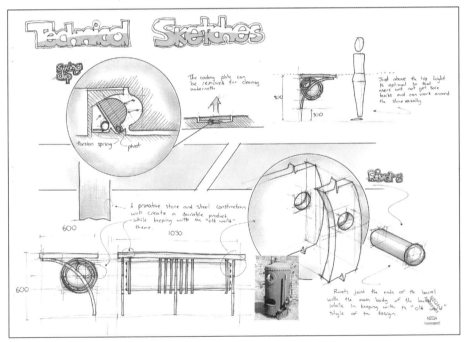

Technical Sketches

The cooking plate can be removed for cleaning underneath

Just above the hip height is optimal so that users will not get sore backs and can work around the stove easily.

torsion spring — pivot

A primative stone and steel construction will create a durable product while keeping with the "old world" theme.

600   1030

600

Rivets joint the ends of the barrel with the main body of the barbecue while in keeping with the "old world" style of the design

ORTHOGRAPHIC PROJECTION

Although the orthographic projection drawings shown are one of two drawings from the brief from which they were taken (*spatial design*), only one drawing is required for assessment at this level.

A second drawing (*product design, not shown*) provides students with two opportunities to achieve the standard, and can show more indepth technical detail in the form of sectioned parts.

The drawings shown should be read *as a guide only*, to give an indication of the step up required at each achievement level.

Orthographic projection drawings should show the following:

- Design features shown accurately drawn
- A minimum of two views in third angle with projection shown
- Visual information not visible in the main outline or complex shape and form
- Title block, indicated scale, labelling, line types, dimensioning
- Neat, precise, clear drawing techniques and conventions

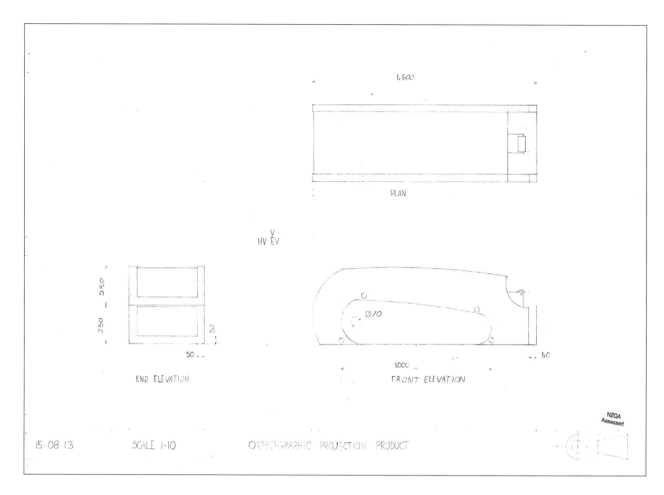

Achieved

CHECKLIST	COMMENT
Design features	Minimal design features shown.
Minimum of two views and correct projection	The student understands orthographic projection, indicated by three views correctly projected from each other.
Visual information not visible in the main outline	An attempt has been made to show information in the form of a sectioned view, but the cutting plane is missing.
Title block, scale, labels, line types, dimensioning	Apart from the end elevation, which is a sectioned view, all have been correctly indicated.
Neat, precise and clear drawing techniques	The overall line quality, and neatness and accuracy of the drawing, needs to be improved.

ISBN: 9780170355575

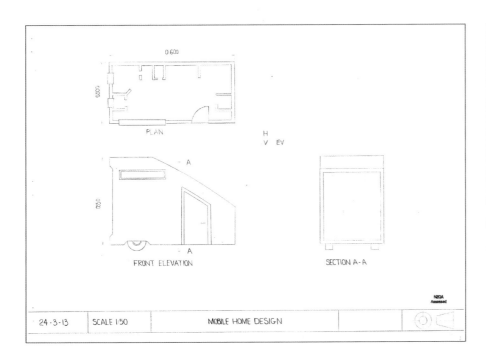

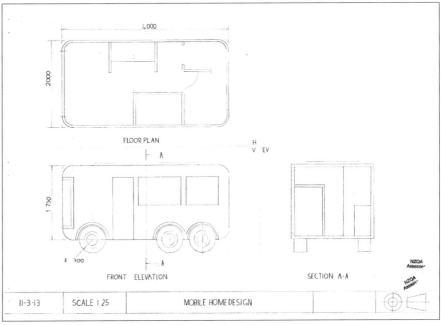

Achieved with merit

CHECKLIST	COMMENT
Design features	A range of the design features is shown.
Minimum of two views and correct projection	The student understands orthographic projection, indicated by three views correctly projected from each other.
Visual information not visible in the main outline	Information in the form of a sectioned view and cutting plane, and a plan view that shows the interior, is clearly seen.
Title block, scale, labels, line types, dimensioning	All have been correctly indicated.
Neat, precise and clear drawing techniques	Overall line quality, and neatness and accuracy of the drawing, is good. It would benefit from sharper line quality at times.

Achieved with excellence

CHECKLIST	COMMENT
Design features	A wide range of the design features is clearly shown.
Minimum of two views and correct projection	The student clearly understands orthographic projection, indicated by three views correctly projected from each other.
Visual information not visible in the main outline	Information in the form of a sectioned view and cutting plane, and a plan view that shows the interior, is clearly seen.
Title block, scale, labels, line types, dimensioning	All have been correctly indicated and well laid out.
Neat, precise and clear drawing techniques	Overall line quality, and neatness and accuracy of the drawing, is precise and clear.

See also page 202

PARALINE DRAWING
Achieved

The paraline drawings shown are of a product design.

To provide students with the opportunity to gain higher grades, a curved surface, circles and/or angles have been incorporated into the designs. This allows for a higher degree of skill to be shown to plot or draw them in 3D.

The drawings shown should be read *as a guide only*, to give an indication of the step up required at each achievement level.

Paraline drawings should show the following:

- Drawings produced using a recognised parallel line pictorial method such as isometric, trimetric, diametric, oblique and planometric
- Drawings are constructed using instrumental drawing techniques
- Construction and outlines are used
- Drawings should comprehensively convey the design intent, using related drawings such as exploded views, sequential views and/or cutaway views
- Drawings include visual information about the internal components, design features and/or complex form
- Precise instrumental drawing techniques such as accuracy in measurement, line intensity and line clarity are used

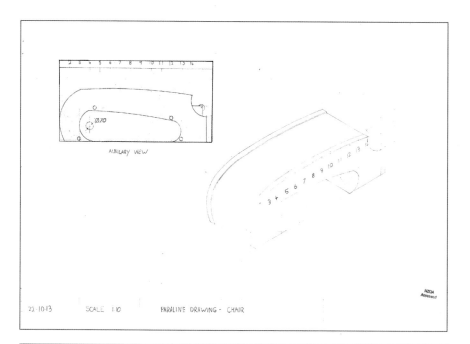

AUXILIARY VIEW

22·10·13   SCALE 1:10   PARALINE DRAWING - CHAIR

CHECKLIST	COMMENTS
Recognised parallel line pictorial drawing method used	The student used, and understands, the isometric pictorial drawing method.
Constructed using instruments	Crating is hard to determine at times. Although evident, it needs to be more clearly defined.
Construction and outlines used	Linework needs to be more defined with a clear difference between construction and outlines.
Exploded, cutaway views, etc. used	The drawing shows a simple exploded cylindrical part. There needs to be more complexity in the drawing for higher grades.
Visual information about the design features shown	There is not enough information about the design shown. The student appears to have reached the limit of his ability.
Precise instrumental drawing techniques used	Drawing techniques need to be more precise. Line intensity, precision, layout, accuracy, etc. all need to be improved to lift the work to a higher level of achievement.

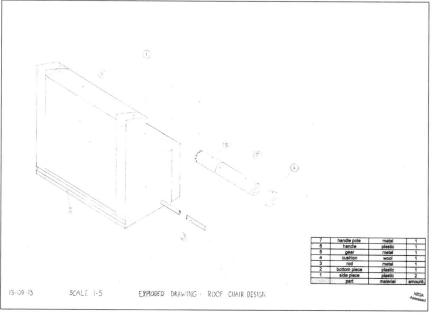

7	handle pole	metal	1
6	handle	plastic	1
5	gear	metal	1
4	cushion	wool	1
3	rod	metal	1
2	bottom piece	plastic	1
1	side piece	plastic	2
part		material	amount

13·09·13   SCALE 1:5   EXPLODED DRAWING - ROOF CHAIR DESIGN

ISBN: 9780170355575   PHOTOCOPYING OF THIS PAGE IS RESTRICTED UNDER LAW.

Achieved with merit

The paraline drawings shown are of a product design.

To provide students with the opportunity to gain higher grades, a curved surface, circles and/or angles have been incorporated into the designs. This allows for a higher degree of skill to be shown to plot or draw them in 3D.

The drawings shown should be read *as a guide only*, to give an indication of the step up required at each achievement level.

Paraline drawings should show the following:

- Drawings produced using a recognised parallel line pictorial method such as isometric, trimetric, diametric, oblique and planometric
- Drawings are constructed using instrumental drawing techniques
- Construction and outlines are used
- Drawings should comprehensively convey the design intent, using related drawings such as exploded views, sequential views and/or cutaway views
- Drawings include visual information about the internal components, design features and/or complex form
- Precise instrumental drawing techniques such as accuracy in measurement, line intensity and line clarity are used

CHECKLIST	COMMENT
Recognised parallel line pictorial drawing method used	The student has understood the isometric pictorial drawing method used. Good use of the auxiliary view to plot the curved end.
Constructed using instruments	Crating, although a little too light in terms of line intensity, is evident, and necessary to draw the views.
Construction and outlines used	Good use of linework with construction and outlines able to be seen.
Exploded, cutaway views, etc. used	Use of simple exploded views shows the construction methods of the main parts of the design.
Visual information about the design features shown	Clear visual information about the design features is shown. The student understands how the parts shown could be constructed.
Precise instrumental drawing techniques used	Consistent application of drawing techniques is evident. Line intensity needs to be a little more defined to lift the work to a higher level of achievement.

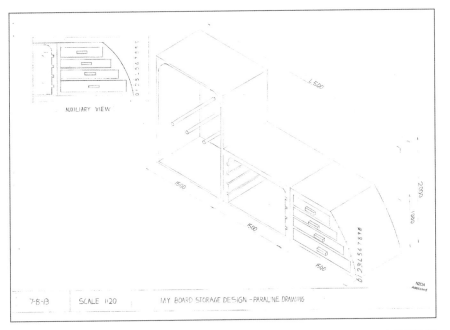

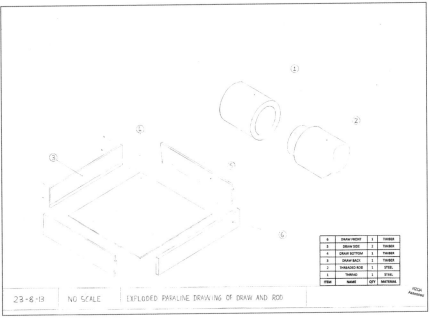

Achieved with excellence

The paraline drawings shown are of a product design.

To provide students with the opportunity to gain higher grades, a curved surface, circles and/or angles have been incorporated into the designs. This allows for a higher degree of skill to be shown to plot or draw them in 3D.

The drawings shown should be read *as a guide only*, to give an indication of the step up required at each achievement level.

Paraline drawings should show the following:

- Drawings produced using a recognised parallel line pictorial method such as isometric, trimetric, diametric, oblique and planometric
- Drawings are constructed using instrumental drawing techniques
- Construction and outlines are used
- Drawings should comprehensively convey the design intent, using related drawings such as exploded views, sequential views and/or cutaway views
- Drawings include visual information about the internal components, design features and/or complex form
- Precise instrumental drawing techniques such as accuracy in measurement, line intensity and line clarity are used

CHECKLIST	COMMENT
Recognised parallel line pictorial drawing method used	The student has clearly used, and understands well, the isometric pictorial drawing method.
Constructed using instruments	Crating is clearly evident, and necessary to draw these complex views.
Construction and outlines used	Superb use of linework with construction and outlines clearly seen.
Exploded, cutaway views, etc. used	Excellent use of exploded views that clearly show the construction methods of the main parts of the design.
Visual information about the design features shown	Clear visual information about the design features is seen across all the drawings. The student understands how the design could be constructed.
Precise instrumental drawing techniques used	Precise and consistent application of drawing techniques clearly evident. Superb use of an auxiliary view to plot the curved surfaces using ordinates.

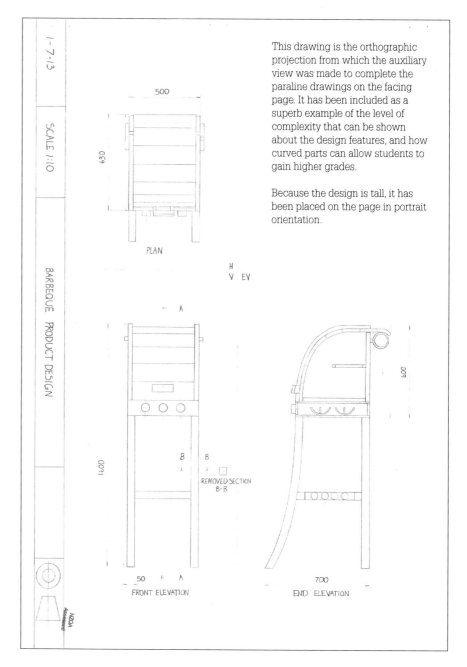

This drawing is the orthographic projection from which the auxiliary view was made to complete the paraline drawings on the facing page. It has been included as a superb example of the level of complexity that can be shown about the design features, and how curved parts can allow students to gain higher grades.

Because the design is tall, it has been placed on the page in portrait orientation.

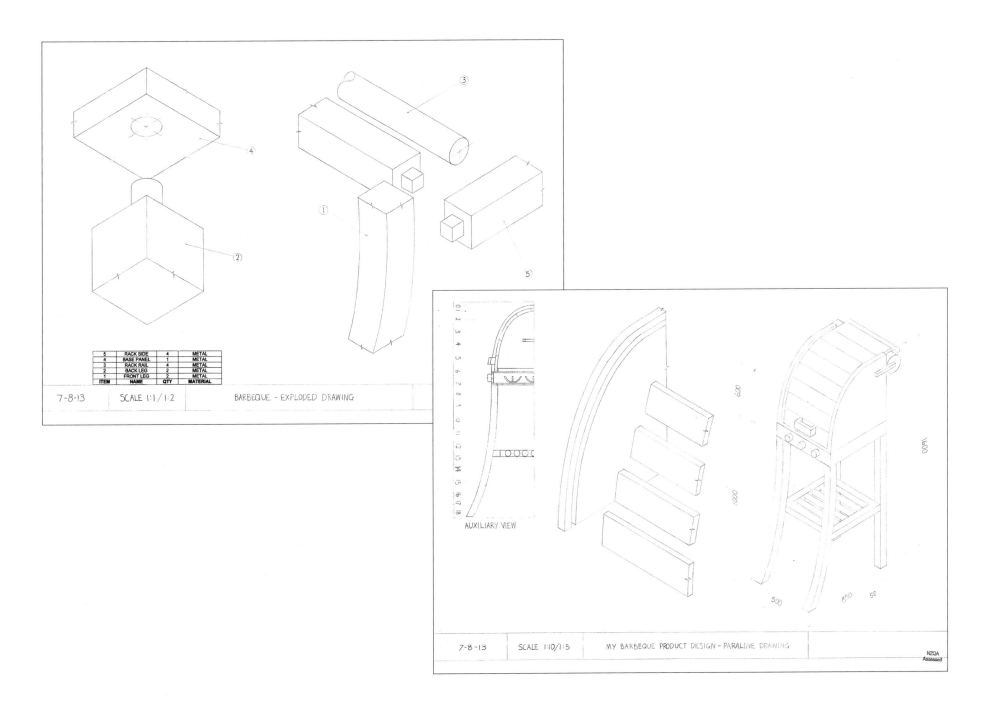

5	RACK SIDE	4	METAL
4	BASE PANEL	1	METAL
3	RACK RAIL	4	METAL
2	BACK LEG	2	METAL
1	FRONT LEG	2	METAL
ITEM	NAME	QTY	MATERIAL

7-8-13	SCALE 1:1 / 1:2	BARBEQUE – EXPLODED DRAWING

AUXILIARY VIEW

7-8-13	SCALE 1:10/1:5	MY BARBEQUE PRODUCT DESIGN – PARALINE DRAWING

NZQA
Assessed

RENDERING TECHNIQUES

Although the rendered drawings shown are of the final design solution from the brief from which they were taken (*spatial and product design*), evidence of rendering techniques should also be shown on ideation pages.

The drawings should be read *as a guide only*, to give an indication of the step up required at each achievement level.

Rendering techniques need to demonstrate the following:

- The tonal qualities produced by an identified light source and its three-dimensional effects on the object's shape and surface qualities
- The consistent and skilful application of rendering techniques to convincingly communicate shape and surface qualtities
- An enhancement of the realistic representation of design qualities to an audience

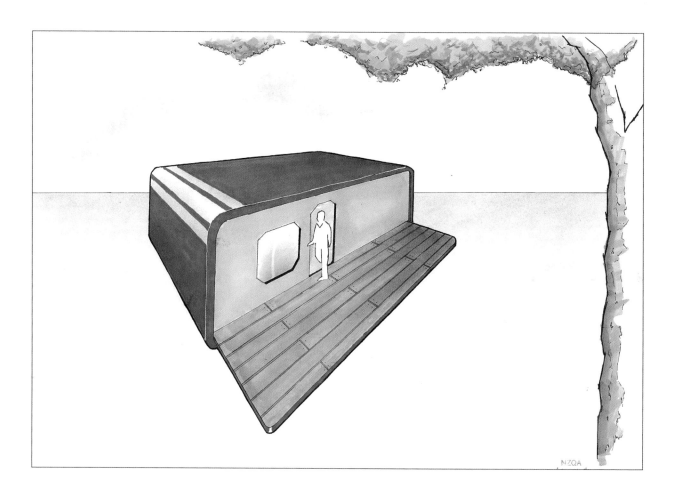

Achieved

CHECKLIST	COMMENT
Identified light source and resulting tonal qualities	Because the top and side have been rendered in the same tone, the light source is not clear.
Rendering techniques communicate shape and surface qualities	The student has attempted rendering techniques, particularly evident on the rounded corner and shine on the window.
Realistic rendering enhances the design qualities to an audience	An attempt has been made to show the design in a realistic form. However, more light and shade, shadows and white highlights would enhance the drawing further.

ISBN: 9780170355575

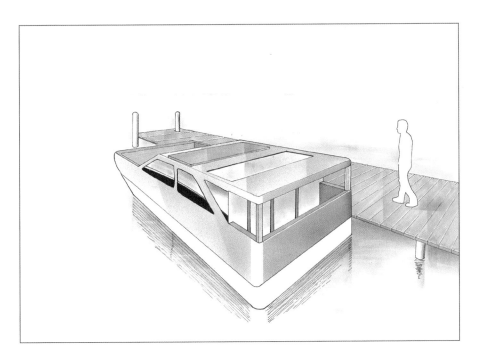

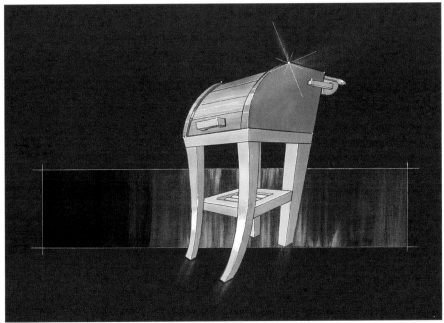

Achieved with merit

CHECKLIST	COMMENT
Identified light source and resulting tonal qualities	A light source is clearly seen by the lighter rendering on the top and darker rendering on the side closest to the viewer.
Rendering techniques communicate shape and surface qualities	The student has attempted rendering techniques well, particularly evident on the rounded corner, the shiny and reflective surface qualities of the wharf, windows, top of the boat and the water.
Realistic rendering enhances the design qualities to an audience	A good attempt has been made to show the design in a realistic form. However, the drawing would benefit from more white highlights on the edges and around the windows.

Achieved with excellence

CHECKLIST	COMMENT
Identified light source and resulting tonal qualities	The light source is from the top left. This is clearly indicated by the star burst and shine on the curved top front parts and the darker surface closest to the viewer.
Rendering techniques communicate shape and surface qualities	Skilled use of rendering techniques and media clearly communicate the shape and surface qualities of the design.
Realistic rendering enhances the design qualities to an audience	The design qualities have been enhanced, not only by skilled rendering, but by careful use of viewpoint and background, providing the drawing with the 'wow' factor.

PERSPECTIVE DRAWING — EXERCISE 1 (PAGE 24)

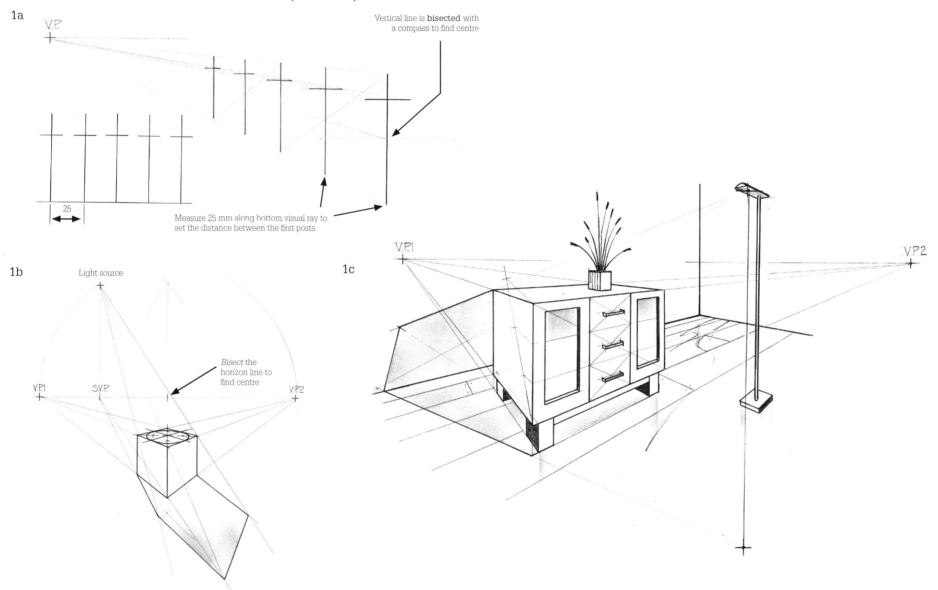

1a

V.P.

Vertical line is **bisected** with a compass to find centre

25

Measure 25 mm along bottom visual ray to set the distance between the first posts

1b

Light source

V.P1   S.V.P.   V.P2

Bisect the horizon line to find centre

1c

V.P1   V.P2

ISBN: 9780170355575   PHOTOCOPYING OF THIS PAGE IS RESTRICTED UNDER LAW.

PERSPECTIVE DRAWING –
EXERCISE 2 AND 3 (PAGE 25–26)

SKILLS

- Constructing equal spaces.
- Locating the furniture correctly, according to their positions in the floor plan.
- Rendering using a range of media.

Notes

- The roman blind has been drawn and rendered separately, then cut and pasted onto the drawing.
- The addition of thin vertical lines heightens the reality of reflection onto the floor and roof.

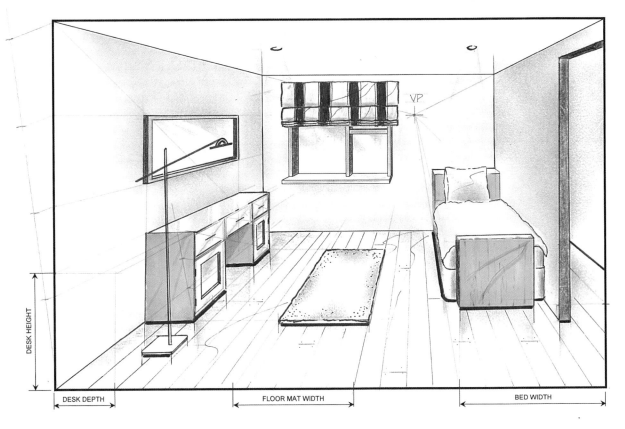

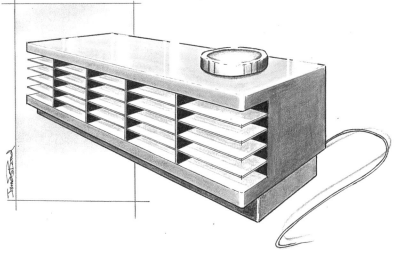

SKILLS

- Constructing equal spaces.
- Working from equal spaces to draw the circular controller.
- Drawing a perspective cylinder (sketched inside a crate).
- Rendering using a range of media.

Notes

- The heater has been rendered using colouring pencils and marker pen.
- Smudged pastel pencil has been used for the background.
- Vertical eraser stripes against an erasing shield create the shiny top.
- The white edges have also been made with an eraser against an erasing shield.

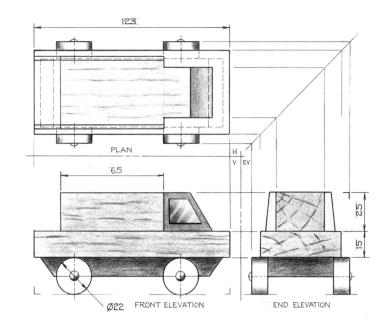

123

PLAN

65

25

15

Ø22   FRONT ELEVATION

END ELEVATION

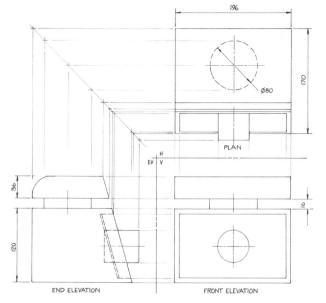

196

170

Ø80

PLAN

36

16

120

END ELEVATION

FRONT ELEVATION

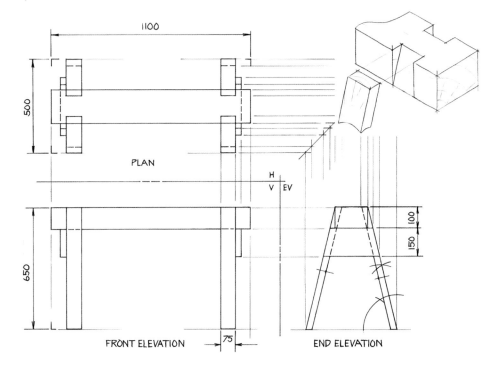

1100

500

PLAN

650

FRONT ELEVATION   75

100

150

END ELEVATION

Construction technique for the end of the builder's saw horse

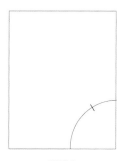

STEP 1
Draw the rectangle to contain the end elevation. Draw a semicircle. From the bottom, mark the radius (60°).

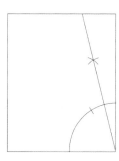

STEP 2
Bisect the distance between the rectangle side and radius mark. Draw a line which will now be at 75°.

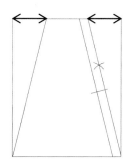

STEP 3
Draw a line at 90° to the 75° angle. Measure along it the leg thickness and draw the inside leg parallel to the outside line. Now draw the left side.

ORTHOGRAPHIC PROJECTION — EXERCISE 5 (PAGE 64)

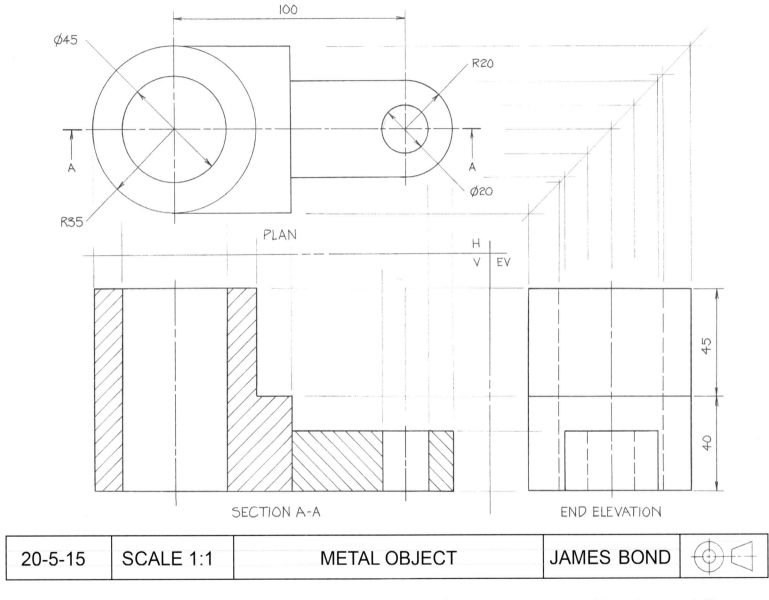

100

Ø45

R20

R35

Ø20

A

A

PLAN

H
V | EV

SECTION A-A

END ELEVATION

45

40

20-5-15	SCALE 1:1	METAL OBJECT	JAMES BOND	

Note: All instrumental drawings should show a title block as above, or similar. (Paraline drawings do not need the projection symbol.)

ISOMETRIC DRAWING — EXERCISES 1, 2, 3 AND 4 (PAGES 69–70)

EXERCISE 1
SKILLS

- Constructing a cylinder and rounded corners.
- Dimensioning. *(Full-size dimensions are shown, not the scaled dimension.)*

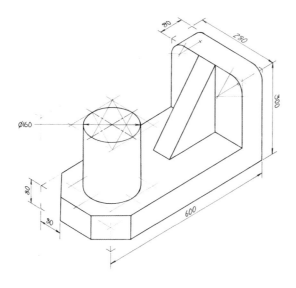

EXERCISE 3
SKILLS

- Drawing an exploded isometric view to show an object pulled apart.
- Using exploded lines to construct the object in line with its adjacent part.
- Rendering to show timber.

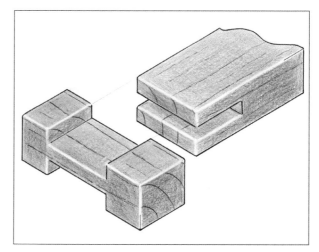

EXERCISE 2
SKILLS

- Constructing a circle and rounded corners.
- Finding the compass points for drawing the recessed circle (the lens) on the end.
- Using an ellipse template for the cylinder on the side.
- Dimensioning. *(Full-size dimensions are shown, not the scaled dimension.)*

The height has been bisected with a compass to find the centre for the ellipse on the side.

Note the compass constructions required for drawing the rounded corners of the bracket design.

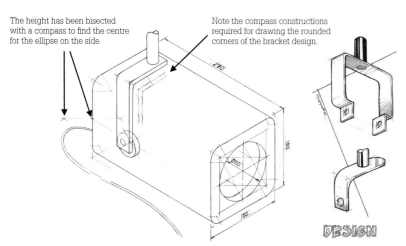

EXERCISE 4
SKILLS

- Drawing an exploded isometric view to show components pulled apart.
- Using exploded lines to construct the object in line with its adjacent part.
- Using an ellipse template to draw circular parts.

Draw the back wheel and axle here

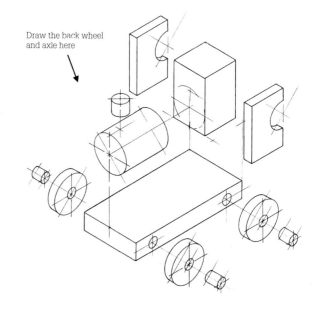

OBLIQUE DRAWING — EXERCISES 1 AND 2 (PAGE 74)

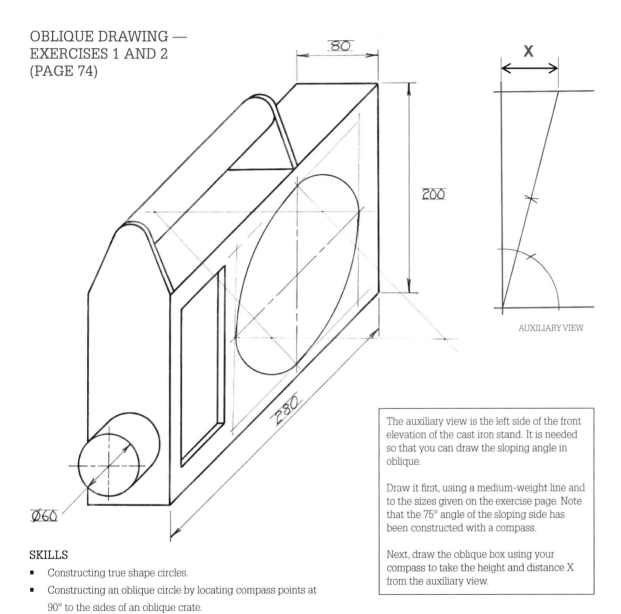

80

200

280

Ø60

X

AUXILIARY VIEW

The auxiliary view is the left side of the front elevation of the cast iron stand. It is needed so that you can draw the sloping angle in oblique.

Draw it first, using a medium-weight line and to the sizes given on the exercise page. Note that the 75° angle of the sloping side has been constructed with a compass.

Next, draw the oblique box using your compass to take the height and distance X from the auxiliary view.

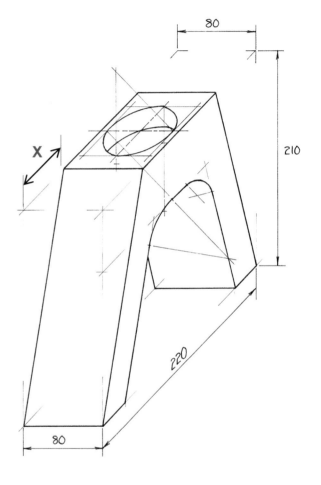

80

210

220

80

X

SKILLS

- Constructing true shape circles.
- Constructing an oblique circle by locating compass points at 90° to the sides of an oblique crate.
- Constructing oblique angles inside a crate.
- Dimensioning (written full size).

SKILLS

- Constructing true shape circles, oblique curves and an oblique circle (a hole).
- Locating compass points for the depth of the hole.
- Constructing 75° with a compass and using an auxiliary view to plot an angle from a 2D to a 3D drawing.
- Dimensioning (written full size).

BUILDING CONSTRUCTION — EXERCISES 1 AND 2
(PAGE 181 AND 182)

EXERCISE 1

EXERCISE 2

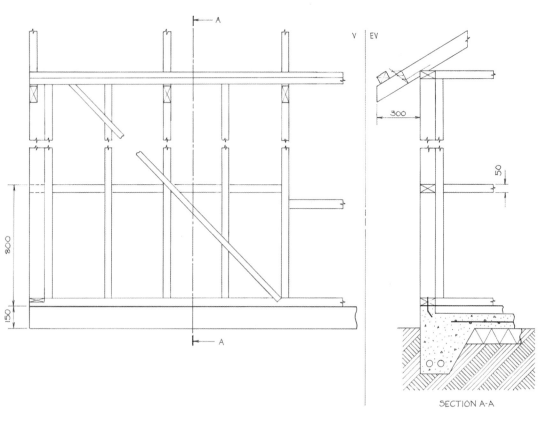

SECTION A-A

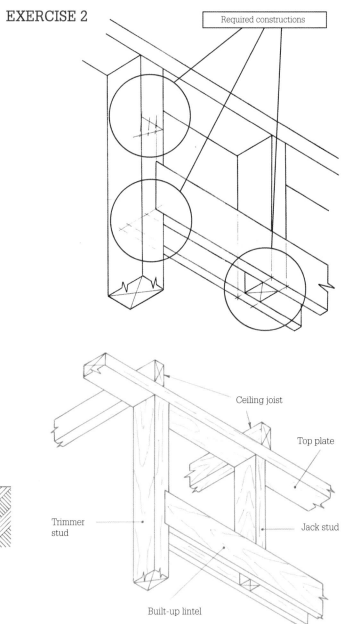

Required constructions

Ceiling joist

Top plate

Jack stud

Trimmer stud

Built-up lintel

ISBN: 9780170355575

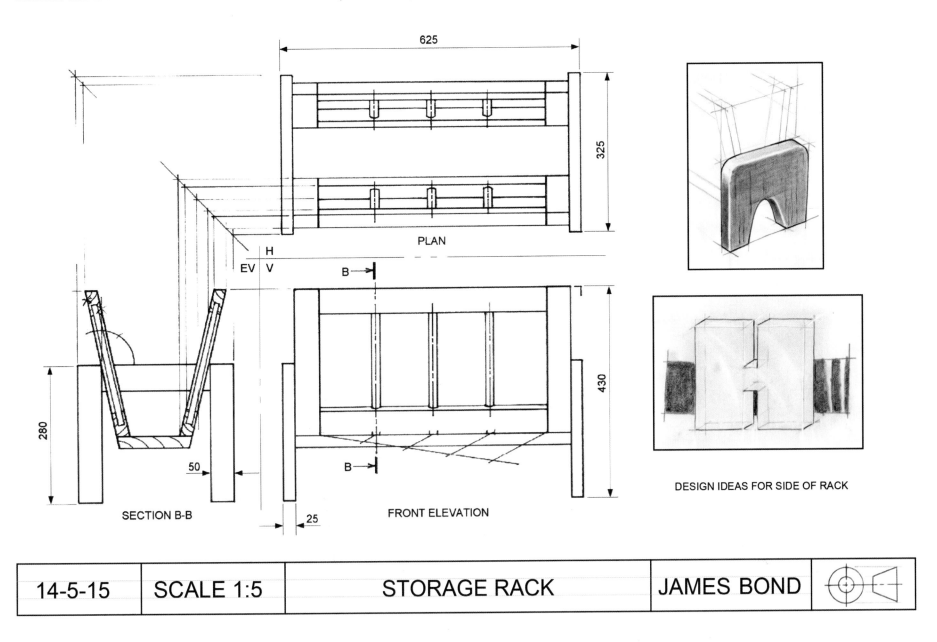

625

325

PLAN

H

EV | V

B →

B →

280

50

SECTION B-B

25

FRONT ELEVATION

430

DESIGN IDEAS FOR SIDE OF RACK

| 14-5-15 | SCALE 1:5 | STORAGE RACK | JAMES BOND | |

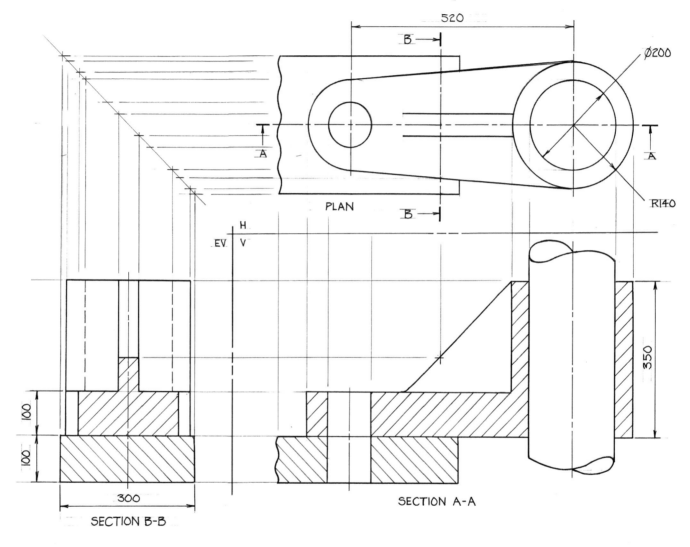

520

B

Ø200

PLAN

B

R140

H
EV V

A

A

350

100

100

300

SECTION B-B

SECTION A-A

| 18-6-15 | SCALE 1:5 | CRANK ARM | JAMES BOND | |

ISBN: 9780170355575   PHOTOCOPYING OF THIS PAGE IS RESTRICTED UNDER LAW.

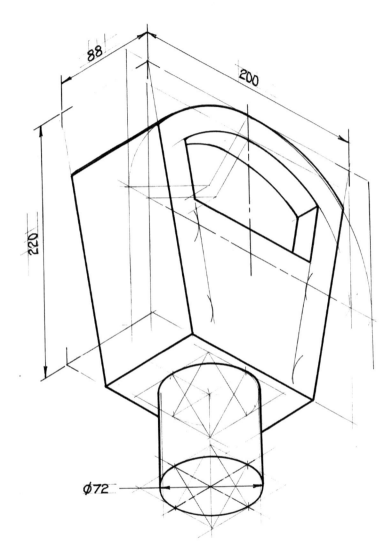

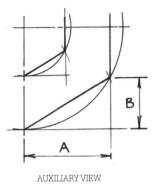

AUXILIARY VIEW

1 Draw an auxiliary view of the two hexagons that form the top and bottom of the lamp shade. (It must be the same size as the isometric drawing, but you do not need the entire hexagon shape.)
2 Place a box about the hexagon shapes.
3 Redraw the full-size boxes in isometric. Start with the centre lines and step out from their centres the distances from the centres in the auxiliary view.
4 Step out the distances A and B from the auxiliary view, along the sides of the isometric boxes, to find where the corners of the lamp will be.

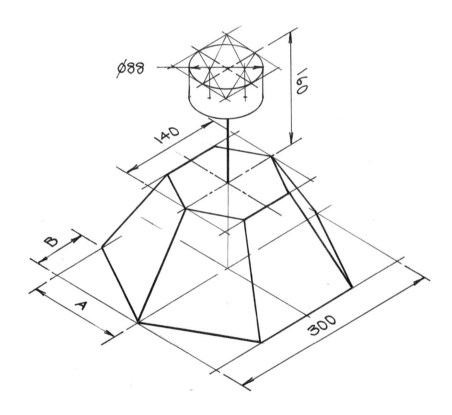

Note the construction of parallel lines using a compass and straight edge to make the sides of the display window parallel to the side of the meter.

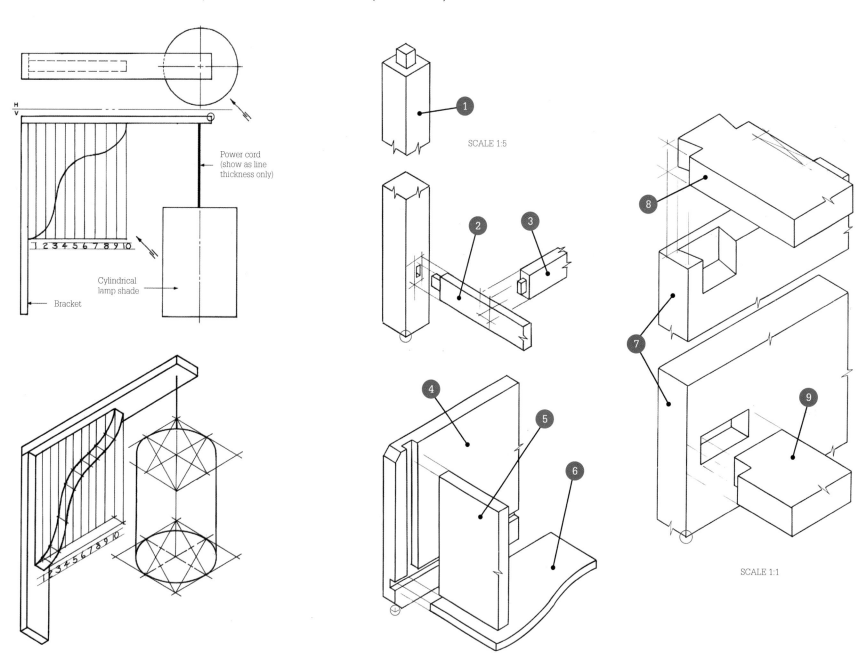

Power cord
(show as line
thickness only)

Cylindrical
lamp shade

Bracket

1 2 3 4 5 6 7 8 9 10

SCALE 1:5

SCALE 1:1

ISBN: 9780170355575

PARALINE DRAWING (OBLIQUE) — EXERCISES 1 AND 2 (PAGES 188 AND 189)

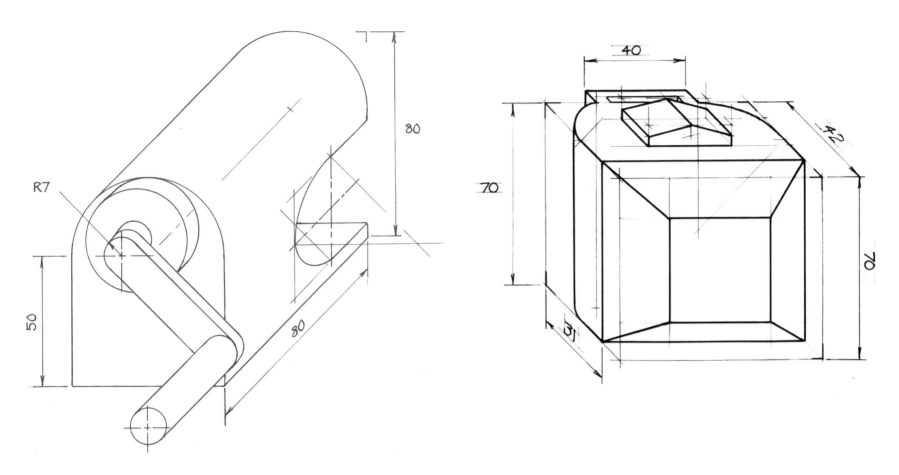

To draw the semicircle on the side of the pencil sharpener, draw the centre lines first, then the box with sides parallel to the centre lines. The distance from the centre to the side of the box will be 20 mm (the radius of the curve).

Compass points for drawing the curve are found by drawing a horizontal line and two 45° lines where the centre lines touch the sides of the box.

PERSPECTIVE DRAWING — EXERCISE 2 (PAGE 191)

Suggested answer

Rendering media: *marker pens, pastel pencil, colouring pencil, eraser, black fineline pens*

Matthew Beneka, Year 11

Rendering media: *marker pens, pastel pencil, colouring pencil, eraser, black fineline pens*

Notes

- Trees in the foreground (New Zealand cabbage tree) and a human figure give the drawing scale.
- Sky is rendered with blue pastel pencil, smudged from the edge of a piece of paper, fading upwards into white space. To make the jagged edge of a cloud effect seen in Matthew's drawing on the right, pastel has been smudged from the edge of a piece of torn paper.
- Drawings should bleed into white space.
- Rendering is lighter in the background, darker in the foreground.
- A shadow of the roof overhang on the walls heightens realism (4B pencil has been smudged from the edge of a piece of paper, parallel to the roof slope).
- Pastel pencil, smudged from the edge of a piece of paper, has been used to render the walls. (Walls opposite the light source are darker.)

- Reflection of light onto the window has be done with a black marker pen.
- Reflection on the deck has been made with an eraser against an erasing shield. (A piece of paper also works.) Note there is a wide and a thin stripe. The wide stripe has one sharp edge that fades towards the opposite side. *The stripes must be vertical.*
- Vertical reflection of the deck edge on the water in Matthew's drawing has been done with a fineline black pen against a straight edge.
- A range of colours has been used on foliage, using a combination of marker pen and colouring pencil.
- Grass has been rendered with a dark- and light-green colouring pencil, using horizontal strokes. It is darker near the edges and under foliage, fading into white space.